Energy-Efficient Computing and Communication

Special Issue Editor

Sangheon Pack

MDPI • Basel • Beijing • Wuhan • Barcelona • Belgrade • Manchester • Tokyo • Cluj • Tianjin

Special Issue Editor
Sangheon Pack
Korea University
Korea

Editorial Office
MDPI
St. Alban-Anlage 66
4052 Basel, Switzerland

This is a reprint of articles from the Special Issue published online in the open access journal *Energies* (ISSN 1996-1073) (available at: https://www.mdpi.com/journal/energies/special_issues/Energy_Computing_Communication).

For citation purposes, cite each article independently as indicated on the article page online and as indicated below:

LastName, A.A.; LastName, B.B.; LastName, C.C. Article Title. *Journal Name* **Year**, *Article Number*, Page Range.

ISBN 978-3-03936-148-9 (Pbk)
ISBN 978-3-03936-149-6 (PDF)

Contents

About the Special Issue Editor

Sangheon Pack received B.S. and Ph.D. degrees from Seoul National University, Seoul, Korea, in 2000 and 2005, respectively, both in computer engineering. In 2007, he joined the faculty of Korea University, Seoul, Korea, where he is currently a Professor in the School of Electrical Engineering. From 2005 to 2006, he was a Postdoctoral Fellow with the Broadband Communications Research Group, University of Waterloo, Waterloo, ON, Canada. He was the recipient of IEEE/Institute of Electronics and Information Engineers (IEIE) Joint Award for IT Young Engineers Award 2017, Korean Institute of Information Scientists and Engineers (KIISE) Young Information Scientist Award 2017, Korea University TechnoComplex (KUTC) Crimson Professor 2015, Korean Institute of Communications and Information Sciences (KICS) Haedong Young Scholar Award 2013, LG Yonam Foundation Overseas Research Professor Program in 2012, and IEEE ComSoc APB Outstanding Young Researcher Award in 2009. He was the head of the Department of Samsung IT Convergence at Korea University in 2012, and he is the Department Head of Hyundai Automotive Convergence in Korea University. In 2003, he was a Visiting Researcher at the Fraunhofer Institute for Open Communication Systems (FOKUS), Berlin, Germany. He served as a TPC vice-chair for information systems of IEEE WCNC 2020, a track chair of IEEE CCNC 2019, a TPC chair of EAI Qshine 2016, a publication co-chair of IEEE INFOCOM 2014 and ACM MobiHoc 2015, a co-chair of IEEE VTC 2010-Fall transportation track, a co-chair of IEEE WCSP 2013 wireless networking symposium, a TPC vice-chair of ICOIN 2013, and a publicity co-chair of IEEE SECON 2012. He is an editor of IEEE Internet of Things (IoT), Journal of Communications Networks (JCN), IET Communications, and he is a Guest Editor of IEEE Transactions on Emerging Topics in Computing (TETC) and IEEE Transactions on Network Science and Engineering (TNSE). He is a senior member of the IEEE. His research interests include softwarized networking (SDN/NFV), 5G/6G mobile core networks, mobile edge computing/ programmable data plane, and vehicular networking.

Preface to "Energy-Efficient Computing and Communication"

Improving the energy efficiency in communications and computing systems has become one of the most important issues to realize green ICT. Even though a number of studies have been conducted, most of them focused on one aspect—either communications or computing systems. However, salient features in communications and computing systems should be jointly considered, and novel holistic approaches across communications and computing systems are required to implement energy-efficient systems. In this regard, this Special Issue aimed to gather recent advances in energy-efficient communications and computing technologies.

Park et al. [1] propose a novel scheme that improves the energy efficiency and network throughputs by controlling the topology of the multi-unmanned aerial vehicle (UAV) network. The use of UAVs has been researched in various industrial fields, and a number of studies on operating multiple autonomous networking UAVs suggest a potential use of UAVs in large-scale environments. However, achieving efficient performance in multi-UAV operations remains challenging in terms of energy efficiency, network overhead, and so on. The proposed network topology control scheme functions between the data link layer (L3) and the network layer (L2), and the proposed methodology includes swarm intelligence, meaning that whole topology control can be achieved with lower cost and effort, and without a centralized controller. The experimental results confirm the improvement in performance of the proposed method compared to previous approaches.

Tran et al. [2] implemented a novel one-bit coding metasurface that is capable of focusing and steering beams for enhancing power transfer efficiency of electromagnetic (EM) wave-based wireless power transfer systems. The proposed metasurface includes 16 × 16 unit cells that were designed with a fractal structure and the operating frequency of 5.8 GHz. By appropriately handling the on/off states of the coding metasurface, the reflected EM wave impinged on the metasurface can be controlled. To verify the working ability of the coding metasurface, a prototype metasurface with a control board was fabricated and measured. The experimental results demonstrate that the coding metasurface is capable of focusing a beam to a desired direction. In addition, for practical scenarios, the authors propose an adaptive optimal phase control scheme for focusing the beam to a mobile target and proved that the proposed adaptive optimal phase control scheme outperforms the random phase control and beam synthesis schemes.

Mobile edge computing (MEC) technology was developed to mitigate the overload problem in networks and cloud systems. An MEC system computes the offloading computation tasks from resource-constrained Internet of Things (IoT) devices. Several convergence technologies with renewable energy resources (RERs) such as photovoltaics have been proposed to improve the survivability of IoT systems. Parck et al. [3] propose an MEC integrated with RER system, denoted energy-harvesting (EH) MEC. Since the energy supply of RERs is unstable forvarious reasons, EH MEC needs to consider the state-of-charge (SoC) of the battery to ensure system stability. Therefore, the authors devised an offloading scheduling algorithm considering the EH MEC battery as well as the service quality of experience (QoE). In the first stage of the scheduling algorithm, a non-convex optimization problem was formulated and a greedy algorithm was constructed to obtain approximate optimal solutions. In the second stage, based on Lyapunov optimization, a low-complexity algorithm is proposed that considers both the workload queue and battery stability.

Ko et al. [4] propose an energy efficient cooperative computation algorithm (EE-CCA), where a pair of IoT devices decides whether to offload some parts of the task to the opponent by considering their energy levels and the task deadline. To minimize the energy outage probability while completing most tasks before their deadlines, a constraint Markov decision process (CMDP) problem is formulated and the optimal offloading strategy is obtained by linear programming (LP). The evaluation results demonstrate that the EE-CCA can reduce the energy outage probability up to 78% compared with the random offloading scheme while completing tasks before their deadlines with high probability.

Ko et al. [4] propose an energy efficient cooperative computation algorithm (EE-CCA), where a pair of IoT devices decides whether to offload some parts of the task to the opponent by considering their energy levels and the task deadline. To minimize the energy outage probability while completing most tasks before their deadlines, a constraint Markov decision process (CMDP) problem is formulated and the optimal offloading strategy is obtained by linear programming (LP). The evaluation results demonstrate that the EE-CCA can reduce the energy outage probability up to 78% compared with the random offloading scheme while completing tasks before their deadlines with high probability.

For energy-neutral operation (ENO) of wireless sensor networks (WSNs), Choi and Lee [6] applied a wireless-powered communication network (WPCN) to a WSN with a hierarchical structure. In this hierarchical WPSN, sensor nodes with high harvesting energies and good link budgets have energy remaining after sending their data to the cluster head (CH), whereas the CH suffers from energy scarcity. The authors applied the simultaneous wireless information and power transfer (SWIPT) technique to the considered WPSN so that the sensor nodes can transfer their remaining energy to the CH while transmitting data in a cooperative manner. To maximize the achievable rate of sensing data while guaranteeing ENO, a novel ENO framework is presented that provides a frame structure for SWIPT operation, rate improvement subject to ENO, SWIPT ratio optimization, as well as clustering and CH selection algorithm.

Joung et al. [6] propose a power control method for a buffer-aided relay node (RN) to enhance the energy efficiency of the RN system. By virtue of a buffer, the RN can reserve the data at the buffer when the the channel gain between an RN and a destination node (DN) is weaker than that between an SN and RN. The RN then opportunistically forwards the reserved data in the buffer according to channel condition between the RN and the DN. By exploiting the buffer, the RN reduces transmit power when it reduces the transmission data rate and reserves the data in the buffer. Therefore, without any total throughput reduction, the power consumption of RN can be reduced, resulting in the energy efficiency (EE) improvement of the RN system. For power control, a simple power control method was devised based on a two-dimensional surface fitting model of an optimal transmit power of RN.

These papers offer a broad view of the relevant, diversified, and challenging problems arising in energy-efficient communications and computing. I would like to express my sincere thanks to all the authors, reviewers, and the staff at MDPI.

Sangheon Pack
Guest Editors

References

1. Park, S.; Kim, H.T.; Kim, H. Energy-Efficient Topology Control for UAV Networks. *Energy* **2019**, *12*, 4523.
2. Tran N.; Amri, M., Park, J.; Hwang, S.; Kim, D.; Choi, K. A Novel Coding Metasurface for Wireless Power Transfer Applications. *Energy* **2019**, *12*, 4488.
3. Park, L.; Lee, C.; Na, W.; Choi, S.; Cho, S. Two-Stage Computation Offloading Scheduling Algorithm for Energy-Harvesting Mobile Edge Computing. *Energy* **2019**, *12*, 4367.
4. Ko, H.; Lee, J.; Jang, S.; Kim, J.; Pack, S. Energy Efficient Cooperative Computation Algorithm in Energy Harvesting Internet of Things. *Energy* **2019**, *12*, 4050.
5. Choi, H.; Lee, J. Energy-Neutral Operation Based on Simultaneous Wireless Information and Power Transfer for Wireless Powered Sensor Networks. *Energy* **2019**, *12*, 3823.
6. Joung, J.; Lee, H.; Zhao, J.; Kang, X. Power Control Method for Energy Efficient Buffer-Aided Relay Systems. *Energy* **2019**, *12*, 3234.

energies

Article

Power Control Method for Energy Efficient Buffer-Aided Relay Systems

Jingon Joung [1], Han Lim Lee [1], Jian Zhao [2] and Xin Kang [3,*]

1 School of Electrical and Electronics Engineering, Chung-Ang University, Seoul 06974, Korea
2 School of Electronic Science and Engineering, Nanjing University, Nanjing 210023, China
3 Center for Intelligent Networking and Communications (CINC), University of Electronic Science and Technology of China (UESTC), Chengdu 611731, China
* Correspondence: kangxin@uestc.edu.cn; Tel.: +82-2-820-5145

Received: 2 June 2019; Accepted: 20 August 2019; Published: 22 August 2019

Abstract: In this paper, a power control method is proposed for a buffer-aided relay node (RN) to enhance the energy efficiency of the RN system. By virtue of a buffer, the RN can reserve the data at the buffer when the the channel gain between an RN and a destination node (DN) is weaker than that between SN and RN. The RN then opportunistically forward the reserved data in the buffer according to channel condition between the RN and the DN. By exploiting the buffer, RN reduces transmit power when it reduces the transmit data rate and reserve the data in the buffer. Therefore, without any total throughput reduction, the power consumption of RN can be reduced, resulting in the energy efficiency (EE) improvement of the RN system. Furthermore, for the power control, we devise a simple power control method based on a two-dimensional surface fitting model of an optimal transmit power of RN. The proposed RN power control method is readily and locally implementable at the RN, and it can significantly improve EE of the RN compared to the fixed power control method and the spectral efficiency based method as verified by the rigorous numerical results.

Keywords: UAV; relay; cooperative communications; buffer; power control; energy efficiency

---

1. Introduction

Initiated by the information theoretical perspective research on the cooperative communications using a relay node (RN) [1–5], various cooperative communication techniques have been studied to improve communication reliability, such as spectral efficiency (SE) and bit-error-rate (BER), and/or enlarge the coverage of the communications [6,7]. For example, in works by the authors of [2,3], SE was considered as a key performance metric of the cooperative networks, and in the work by the authors of [4], source precoder, relaying matrices, and destination decoder were iteratively optimized to improve BER performance. There are various retransmission strategies and duplex methods for the relaying. Amplify-and-forward (AF) relays simply amplify and forward (retransmit) the received signal [8–12], while decode-and-forward (DF) relays decode, encode, and forward the signal under certain conditions [2,13]. Full-duplex (FD) relay can transmit and receive the signal simultaneously, while a half-duplex (HD) relay receives and transmits separately [2].

Recently, unmanned aerial vehicles (UAV) have been rigorously studied as RN or base station (BS) [14–17]. If the UAV is employed as a relay, i.e., RN, it can forward the ground BS signals to other UAVs in the air. Here, a highly limited resource, i.e., energy, should be carefully managed to prolong the battery life of the RN. To this end, energy efficiency (EE) of the cooperative system has to be emphasized in resource management. For example, in the work by the authors of [18], a resource allocation strategy including power allocation was studied for a relay system to enhance the EE of the cooperative system. In the work by the authors of [19], a near-optimal iterative subcarrier pairing algorithm and power allocation was proposed to improve EE of DF relay networks. For a mobile RN, SE and EE are considered [20].

Based on the observation that the same achievable rate can be achieved with lower transmit power of the RN compared to an RN without the buffer, it was shown that the achievable rate can be increased by employing a buffer at the RN (see works by the authors of [21–24] and the references therein). In the work by the authors of [24], the frequency and power resources were optimally allocated for multiple nodes to improve the EE of the buffer-aided relay networks. Adapting the buffer at RN allows the opportunistic retransmission, and the average achievable rate performance can thus be improved. Moreover, the packet delay and outage probability can also be reduced by using the buffer [25].

In this study, the EE of a buffer-aided fixed RN system is investigated under the assumption that the transmit power of the source node (SN) is also fixed. Here, we assume that SN and HD-DF RN are located at the fixed positions as shown in Figure 1. The RN serves the SN and destination node (DN) as a signal repeater that forwards the received signals from the SN to DN. This scenario is similar to the conventional cellular communication scenario using a repeater that forwards the outdoor base station (i.e., SN) signals to an indoor mobile user (i.e., DN) in a building [26]. Thus, the assumption of the fixed SN and RN are reasonable and many of studies regarding the cooperative communications has been performed for the fixed RN [27–30]. For the buffer-aided fixed RN, we propose a method to control the transmit power of RN based on the RN power consumption model and the distance between RN and DN to improve the network EE. The achievable rate is derived, and using it and an RN power consumption model, the EE of the network is formulated. Since it is intractable to derive the derivative of the EE function with respect to the transmit power and EE is a function of various power consumption parameters, analytical design of the optimal power control strategy is formidable. Since analytical optimization methods, e.g., the resource allocation method in the work by the authors of [24], require the channel state information among the nodes at the transmitter, the practical implementation is challenging. Even any conventional one-dimensional numerical search algorithm or a learning-based method, e.g., a reinforcement learning, is not applicable as it requires the achievable rate at DN at each iteration. This causes a significant network overhead as the achievable rate information or channel state information is needed to be fed back from DN to RN. Therefore, we devise a simple power control method based on a model of the optimal transmit power of RN. To model the optimal transmit power of RN, we extract two essential parameters that determine the transmit power of RN: (i) constant power consumption P_c at RN and (ii) distance d_{rd} between RN and DN. The optimal transmit power of RN are then modeled with respect to P_c and d_{rd} by using a two-dimensional surface fitting method. For the RN power control, thus, only P_c and d_{rd} information is required at RN. Here, P_c is local information, which can be readily measured at RN. The distance d_{rd} can also be measured at RN based on the received signal strength from DN. Therefore, the proposed RN power control method is easily implementable at the RN. From the rigorous numerical results, it is verified that the proposed power control method can significantly improve EE compared to the fixed power control method regardless of P_c and d_{rd}.

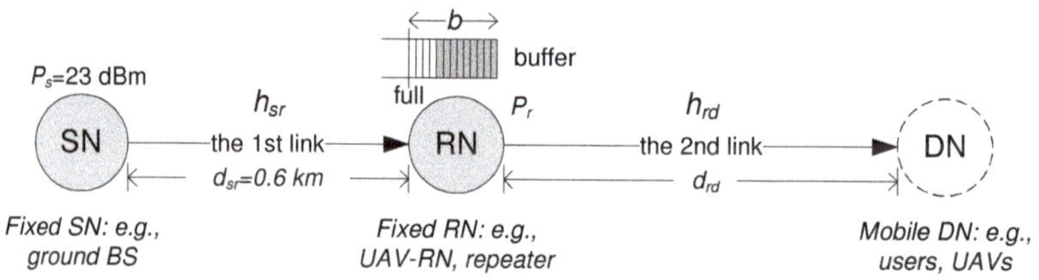

Figure 1. Buffer-aided relay system model, where P_s and P_r are the transmit power of the source node (SN) and relay node (RN), respectively; d_{sr} and d_{rd} are the distance between SN and RN and between RN and DN, respectively; and b is the size of the buffer.

2. RN System Model

Consider an RN system as shown in Figure 1, in which SN transmits data to DN through RN. RN is located at the fixed position d_{sr} km apart from SN. Each node has a single antenna. DN could be either an UAV in the air or ground user, whose direct link to SN is blocked due to the obstacles between SN and DN. The channel of the first link between SN and RN is denoted by $h_{sr} = \sqrt{\rho_{sr}}\bar{h}_{sr}$. Here, ρ_{sr} is the large-scale fading between SN and RN, and small-scale fading $\bar{h}_{sr}$ is independent and identically distributed (i.i.d.) random variables with $\mathcal{CN}(0,1)$ distribution, i.e., Rayleigh fading channels. Similarly, the channel of the second link is modeled as $h_{rd} = \sqrt{\rho_{rd}}\bar{h}_{rd}$, where ρ_{rd} is the large-scale fading between RN and DN and $\bar{h}_{rd} \sim \mathcal{CN}(0,1)$ is the small-scale fading.

At time t, instantaneous achievable rates of the first and second links are written as follows.

$$R_s(t) = \frac{1}{2}\log_2\left(1+\gamma_{sr}(t)\right) = \frac{1}{2}\log_2\left(1+\frac{P_s\rho_{sr}|h_{sr}(t)|^2}{\sigma_r^2}\right), \tag{1}$$

$$R_r(t) = \frac{1}{2}\log_2\left(1+\gamma_{rd}(t)\right) = \frac{1}{2}\log_2\left(1+\frac{P_r\rho_{rd}|h_{rd}(t)|^2}{\sigma_d^2}\right), \tag{2}$$

where $\gamma_{sr}(t)$ and $\gamma_{rd}(t)$ are the instantaneous signal-to-noise ratios (SNRs) at RN and DN, respectively; P_s and P_r are the transmit power of SN and RN, respectively; and σ_r^2 and σ_d^2 are the variances of the additive white Gaussian noise (AWGN) at the RN and DN, respectively; without loss of generality, the average power of the symbols transmitted from SN and RN is assumed to be one.

When the first link is better than the second link, i.e., $\gamma_{sr} \geq \gamma_{rd}$ and equivalently $R_r(t) \geq R_s(t)$, no matter how much information is delivered from SN to RN, the RN can forward no greater than $R_r(t)$. On the other hand, even if $\rho_{sr} < \rho_{rd}$, the RN cannot forward more information than what RN received, i.e., information causality [1,21]. Thus, the instantaneous achievable rate of the overall link at time t is written as follows [2,3]:

$$R(t) = \min\{R_s(t), R_r(t)\}. \tag{3}$$

Using (3), the average achievable rate for T seconds is obtained as

$$R(P_r) = \frac{1}{T}\sum_{t=1}^{T} R(t), \tag{4}$$

where note that the average achievable rate is a function of the transmit power of RN, i.e., P_r, which will be designed later.

3. Achievable Rate of RN with a Buffer

An RN employs a buffer that can reserve the information bits. The buffered bits are received from SN, yet not forwarded to DN if the channel condition of the second link is poor. The bits in the buffer are forwarded later once the channel condition changes to be good. Considering the limited buffer size, b, and information causality, we can consider two cases as follows.

- $R_s(t) \geq R_r(t)$: RN forwards $R_r(t)$ and reserves the remaining bits, i.e., $R_s(t) - R_r(t)$ bits, at the buffer unless the buffer is full with b bits. Thus, the buffer status at time t will be $\min\left(B(t-1) + (R_s(t) - R_r(t)),\, b\right)$, where $B(0) = 0$.
- $R_s(t) < R_r(t)$: RN forwards $R_s(t)$. In this case, since RN can forward more bits up to $R_r(t) - R_s(t)$, the number of forwarded bits will be $R_s(t) + \min\left(B(t-1),\, R_r(t) - R_s(t)\right)$. Accordingly, the buffer status will be $\max\left(B(t-1) - (R_r(t) - R_s(t)), 0\right)$.

Concretely, by virtue of the buffer, the achievable rate of the overall link with the buffer-aided RN and the buffer status at time t are written as follows.

$$R_B(t) = \begin{cases} R_r(t), & \text{if } R_s(t) \geq R_r(t) \\ R_s(t) + \min\left(B(t-1), R_r(t) - R_s(t)\right), & \text{otherwise} \end{cases} \tag{5}$$

$$B(t) = \begin{cases} \min\left(B(t-1) + (R_s(t) - R_r(t)), b\right), & \text{if } R_s(t) \geq R_r(t) \\ \max\left(B(t-1) - (R_r(t) - R_s(t)), 0\right), & \text{otherwise} \end{cases} \tag{6}$$

For example, Figure 2 shows the status of a buffer with $b = 2$ for $T = 500$ when $\gamma_{sr} = 10$ dB. From the results, it is clearly observed that the buffer operates as expected. When the first link is better than the second link, i.e., $\gamma_{sr} \gg \gamma_{rd}$, the information amount delivered from SN to RN is greater than that forwarded from RN to DN. Thus, the buffer is almost always full when $\gamma_{rd} = 0$ dB, as shown in Figure 2a. On the other hand, as γ_{rd} increases to 10 dB and 20 dB, the information amount forwarded from RN to DN increases, resulting in the reduction of information bits in the buffer as shown in Figure 2b,c, respectively. When $\gamma_{rd} = 30$ dB, the buffer is almost empty as shown in Figure 2d. Here, we propose that the size of the buffer is critical for the achievable rate of the RN networks.

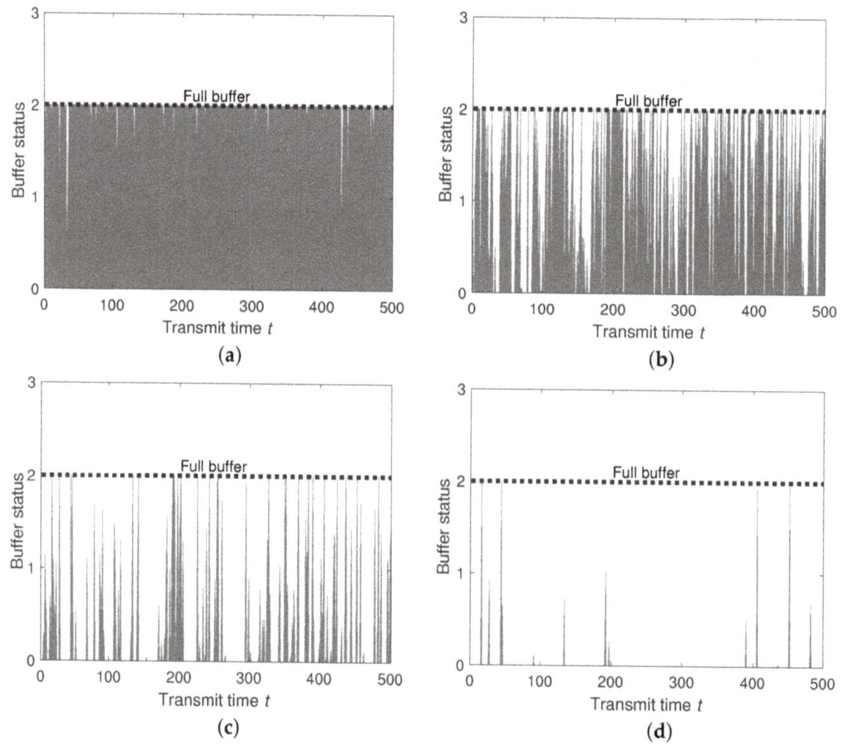

Figure 2. Buffer status when $\gamma_{sr} = 10$ dB and buffer size is two bits, i.e., $b = 2$. (**a**) $\gamma_{rd} = 0$ dB. (**b**) $\gamma_{rd} = 10$ dB. (**c**) $\gamma_{rd} = 15$ dB. (**d**) $\gamma_{rd} = 25$ dB.

4. Buffer Size Design

To determine an efficient buffer size, we evaluate the average achievable rate of $R_B(t)$ in Equation (5), which is defined as

$$R_B(P_r) = \frac{1}{T}\sum_{t=1}^{T} R_B(t). \tag{7}$$

Figure 3 shows the average achievable rates in Equations (4) and (7) by varying the buffer size b when $\gamma_{sr} = 20$ dB. Data transmission time is given as $T = 10^5$. Obviously, the relay system without a buffer is a special case of the relay system with a buffer whose size is zero, i.e., $b = 0$. By comparing the rates at $b = 0$ and $b > 0$, it is clearly observed that the buffer can improve the achievable rate of the cooperative systems. It is also observed that the achievable rate increases and it is saturated as buffer size b increases. As observed in the results, the average achievable rate is almost saturated when the buffer size $b = 10$, regardless of γ_{rd}. From the results, the size of an effective buffer is determined by ten for the considered relay system in this study.

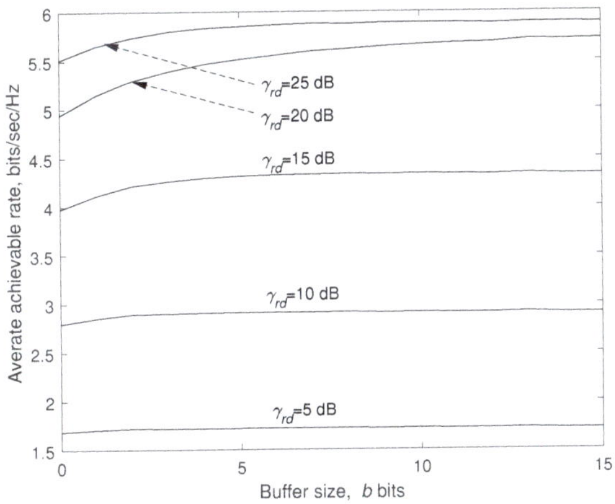

Figure 3. Average achievable rate R_B in Equation (7) over the buffer size b bits when $\gamma_{sd} = 20$ dB.

Now, to justify that the proposed buffer-aided relay system with $b = 10$ outperforms the conventional relay system without a buffer, the average achievable rates $R(P_r)$ and $R_B(P_r)$ defined in Equations (4) and (7), respectively, are evaluated over the transmit power of RN, i.e., P_r.

In Figure 4, $R(P_r)$ in Equation (4) and $R_B(P_r)$ in Equation (7) are compared across P_r. In simulation, it is assumed that the channel attenuation follows the model in works by the authors of [31–34] as follows

$$\rho_A = G - 128 + 10\log_{10}(d_A^{-\alpha}) \text{ dB}, \tag{8}$$

where $A \in \{sr, rd\}$, G includes the transceiver feeder loss and antenna gains, $d_A^{-\alpha}$ is the path loss where d_A is the distance in kilometer between nodes, and α is a path loss exponent. The transmit power of SN is fixed as $P_s = 23$ dBm. In model (8), we set $G = 5$ dB (2 dB and 0 dB feeder losses at the transmitter and receiver, respectively; and 7 dBi and 0 dBi gains for the transmit and receive antennas, respectively [31]), $\alpha = 3.76$ (for urban or suburban environment [35]), $\sigma_r^2 = \sigma_d^2 = -174$ dBm/Hz for AWGN power [31], and $P_s = 23$ dBm for small-size BS [34]. On the other hand, the transmit power of RN varies between 17 dBm and 33 dBm (note that 33 dBm, 21 dBm, and 17 dBm for the transmit power of micro, pico, and femto BSs, respectively [34].). The distance between GBS and RN is set as $d_{sr} = 0.6$ km.

From the results, it is clearly shown that $R_B(P_r) \geq R(P_r)$. For given P_s, the average data rate increases up to a certain point and is saturated as P_r increases. The saturation point of P_r of the buffer-aided relay is lower than that of the relay without a buffer. From this, it is verified that the buffer-aided relay system efficiently utilizes the second-link channel in virtue of opportunistic forwarding; therefore, the greater average achievable rate is achieved. Thus, from this fact, the transmit power of RN can be reduced sustaining the throughput such that it is identical to the throughput without a buffer, i.e., $R_B(P_r^o) = R(P_r)$, where $P_r^o < P_r$. Consequently, it is expected that EE of the network can be improved.

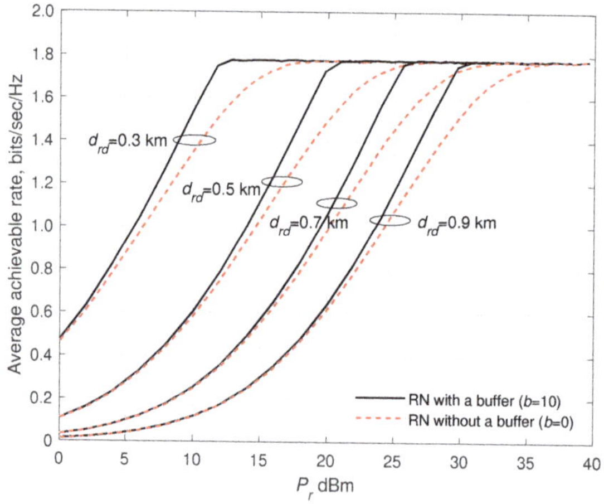

Figure 4. Average achievable rate when $P_s = 23$ dBm, $b = 10$, and $d_{sr} = 0.6$ km.

5. Proposed Power Control Method for Energy Efficient RN

The EE of the network with respect to RN is defined as a ratio of the average achievable rate and the power consumption at RN as follows [32–34,36–38].

$$\mathsf{EE}(P_r) = \frac{R_B(P_r)}{\eta P_r + P_c}, \quad (9)$$

where η represents system inefficiency ($\eta > 1$) that is caused by overhead power consumption at radio frequency circuits, and P_c is the power consumption which is independent of transmit power. The first term in the denominator of Equation (9) is thus the power consumption at the power amplifier of RN. On the other hand, the second term in the denominator of Equation (9) includes a part of power consumption for communication at, for example, a power supply, an alternating current to direct current (AC/DC) converter, a DC/DC converter, and an active cooling system, and the propulsion power consumption for hovering [39].

To design P_r, such that EE in Equation (9) is maximized, we evaluate the EE over P_r. In Figure 5, the EEs of two relay systems with and without a 10-bit buffer are compared when $P_s = 23$ dBm, $d_{sr} = 0.6$ km, and $d_{rd} \in \{0.3, 0.6, 0.9\}$. For the power consumption parameters, we set them as $\eta = 5.26$ and $P_c \in \{10, 20, 30, 40\}$ dBm. Here, the values of simulation parameter are typical for the wireless communication systems (refer to works by the authors of [32–34,36–38] and references therein), and they can be adjusted according to the application systems. Note the the proposed power control framework, which is introduced shortly, is independent of the values of parameters. From the results in Figure 5, we can verify that the EE can be improved by using a buffer at RN, regardless of d_{rd} and P_c.

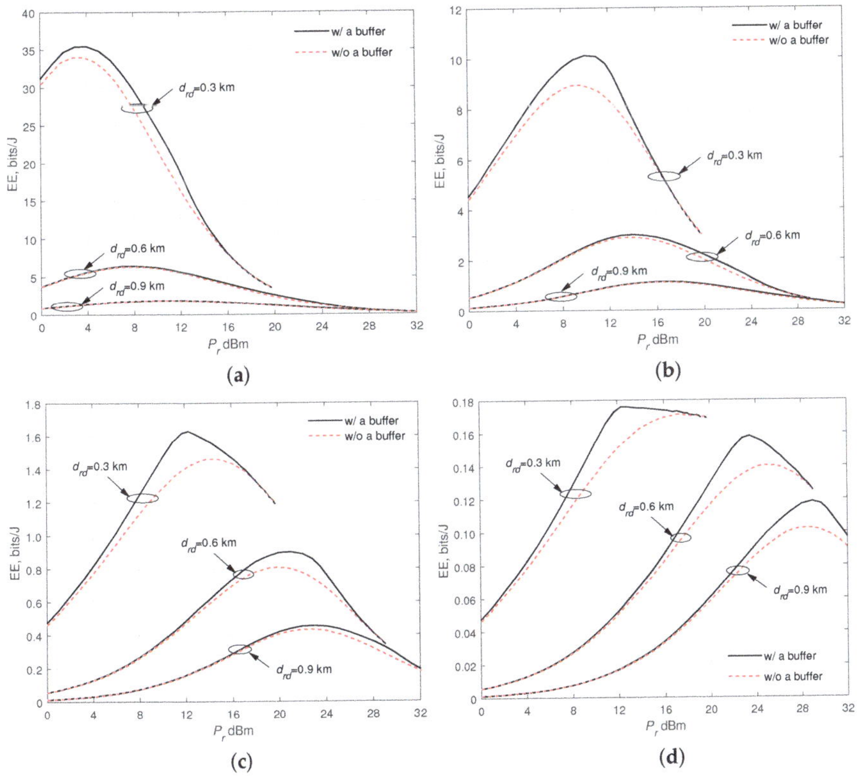

Figure 5. EE across P_r when $P_s = 23$ dBm, $d_{sr} = 0.6$ km, $d_{rd} \in \{0.3, 0.6, 0.9\}$ km, and $\eta = 5.26$. (**a**) $P_c = 10$ dBm. (**b**) $P_c = 20$ dBm. (**c**) $P_c = 30$ dBm. (**d**) $P_c = 40$ dBm.

It is also observed from Figure 5 that the EE function is a unimodal function having a unique maximum of EE. However, to analytically find the optimal P_r maximizing EE is challenging. Using Equations (1), (5), and (7)–(9), we see that the EE function is intractable, i.e., there is no closed form of the first derivative of EE with respect to P_r. One can immediately consider a one-dimensional numerical search algorithms, such as golden section search, quadratic interpolation method, and inexact line searches, to find the local optimal solution [40], or machine learning based algorithm, e.g., reinforcement learning [41–43]. However, since the average achievable rate needs to be fed back from DN to RN for each iteration with adapted P_r to estimate the EE, the iterative approaches require significant overhead of the networks. In the work by the authors of [24], the EE is analytically maximized by optimally allocating the frequency and transmit power resources. To this end, however, the transmitter, namely SN and RN, should know the channel state information among the nodes. Though the analytical approach can provide the optimal EE performance, this analytical strategy is challenging to be practically implemented, due to the network overhead. Thus, in this study, we employ a two-dimensional surface fitting method to model the optimal transmit power of RN as a function of two essential variables of EE, namely d_{rd} and P_c.

In Figure 6, the optimal P_r^o's are shown for $d_{rd} \in \{0.3, 0.6., 0.9\}$ and $P_c \in \{10, 20, 40, 40\}$, which are obtained from Figure 5. The optimal transmit power of RN, i.e., P_r^o increases as P_c or d_{rd} increases. To fit the optimal points of P_r^o to a surface, the multidimensional regression methods are employed [44]. In this study, we employ a polynomial surface fitting method, which is simple to design based on the least squares method and requires a small amount of memory at the RN. The quadratic polynomial surface is modeled as follows.

$$P_r(d_{rd}, P_c) = -8.462 + 27.89d_{rd} + 0.5436P_c - 21.9d_{rd}^2 + 0.5985d_{rd}P_c - 0.008475P_c^2 \quad (10)$$

and it is shown in Figure 7. Similarly, for the optimal points, P_r^o's of RN without a buffer in Figure 6 can be modeled as a surface. The quadratic polynomial surface of P_r^o of RN without a buffer can be modeled as

$$P_r(d_{rd}, P_c) = -8.12 + 25.1d_{rd} + 0.5353P_c - 15.29d_{rd}^2 + 0.2674d_{rd}P_c - 0.003055P_c^2. \quad (11)$$

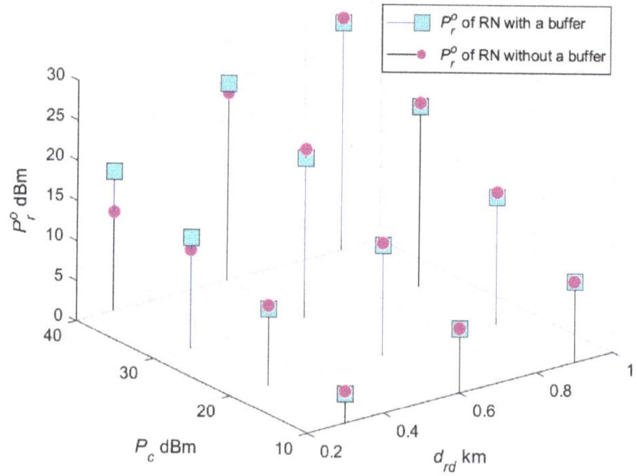

Figure 6. Optimal power P_r^o of RN with and without a buffer over d_{rd} and P_c, where P_r^o is obtained from the results in Figure 5.

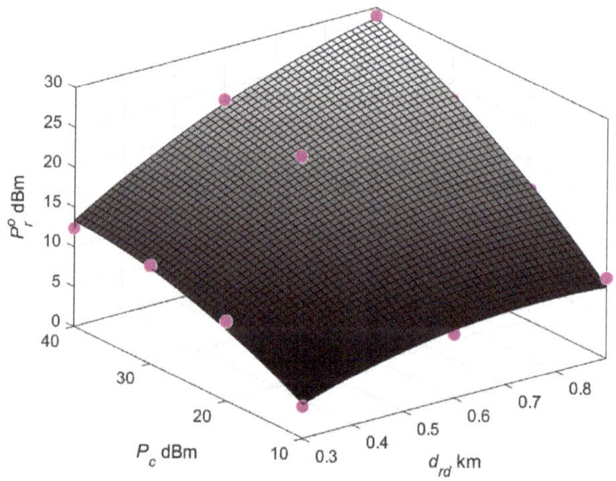

Figure 7. Surface fitted from P_r^o with a buffer in Figure 6 by using a polynomial surface fitting method.

Using Equations (10) and (11), the RN can determine its transmit power P_r if it knows P_c and d_{rd}. Here, it is worth noting that P_c is local information, which can be readily measured at the RN, and that the distance d_{rd} can also be measured at the RN based on the received signal strength from DN. Therefore, the proposed RN power control method can be easily implementable at the RN. For the applications of discrete-level power control, the quantized output of the quadratic polynomial surfaces in Equations (10) and (11) can be used. In this study, we leave the measurement or prediction method for P_c and d_{rd} as future work. In the next section, the proposed RN power control methods using Equations (10) and (11) are verified by evaluating the EE performance.

6. Numerical Results and Discussion

In this section, EE is evaluated by varying P_c and d_{rd}. We compare five power control (PC) schemes:

- EE-based PC w/ a buffer: The proposed RN system with a buffer. The RN transmit power is controlled by a one-dimensional numerical search. This method can be interpreted as a numerical approach of the optimal strategy in the work by the authors of [24], for a single pair of RN and DN. This is optimal yet impractical approach as stated in the previous section.
- EE-based PC w/ a buffer: The proposed RN system with a buffer. The RN transmit power is controlled by Equation (10).
- EE-based PC w/o a buffer: The conventional RN system without a buffer. The RN transmit power is controlled by Equation (11). The PC method is designed in this study, yet it is employed to the conventional RN for the sake of comparison. This method can be interpreted as an EE-based power control method in [18].
- SE-based PC w/ a buffer: PC is performed to maximize SE [2,3]. Here, the RN employs a buffer.
- SE-based PC w/o a buffer: PC is performed to maximize SE [2,3]. Here, the RN does not employ a buffer.

In Figure 8, each subfigure shows EE by varying P_c for fixed d_{rd}. From the results, the proposed PC is verified that it can achieve the near-optimal EE regardless of the power consumption model, i.e., P_c, and the distance d_{rd}. Clearly, EE decreases as P_c increases. It is also observed that the proposed PC can also significantly improve EE for the relay system without a buffer. On the other hand, SE-based PC methods do not provide comparable EE to the relay system with EE-based PC. Here, note that the transmit power of RN of SE-based PC methods is fixed as it is independent of the power consumption model. The fixed parameters of power are as follows: $P_r = 13$ dBm and $P_r = 20$ dBm for RN with and without a buffer, respectively, when $d_{rd} = 0.3$ km in Figure 8a; $P_r = 22$ dBm and $P_r = 30$ dBm for RN with and without a buffer, respectively, when $d_{rd} = 0.5$ km in Figure 8b; $P_r = 27$ dBm and $P_r = 35$ dBm for RN with and without a buffer, respectively, when $d_{rd} = 0.7$ km in Figure 8c; and $P_r = 32$ dBm and $P_r = 40$ dBm for RN with and without a buffer, respectively, when $d_{rd} = 0.9$ km in Figure 8d. This is because, as previously shown in Figure 6, the transmit power of RN is needed to increase to improve EE as P_c or d_{rd} increases. The SE-based fixed PC may achieve the best EE, e.g., when P_c is high. However, the fixed PC strategies are inefficient since P_c may vary in time due to the external environment variation, such as temperature and humidity. Moreover, even P_c is fixed, d_{rd} may vary and accordingly the optimal P_r also varies. To clarify it, the next simulation is performed for fixed P_c by varying d_{rd}.

In Figure 9, each subfigure shows EE by varying d_{rd} for fixed P_c. Clearly, EE decreases as d_{rd} increases. The proposed PC can achieve the near-optimal EE regardless of the power consumption model, i.e., P_c, and the distance d_{rd}. It is also observed that the proposed EE-based PC can also improve EE for the RN system without a buffer, especially, when P_c is small. However, the EE gap between RN with and without a buffer increases as P_c increases. From this results, we see that the buffer is more efficient when P_c is large. Though the SE-based PC methods optimize P_r according to d_{rd}, their EE performance is not comparable to the EE of the proposed EE-based PC methods for RN with a buffer. As expected, the proposed RN system with a buffer and PC is superior to others regardless of P_c and d_{rd}.

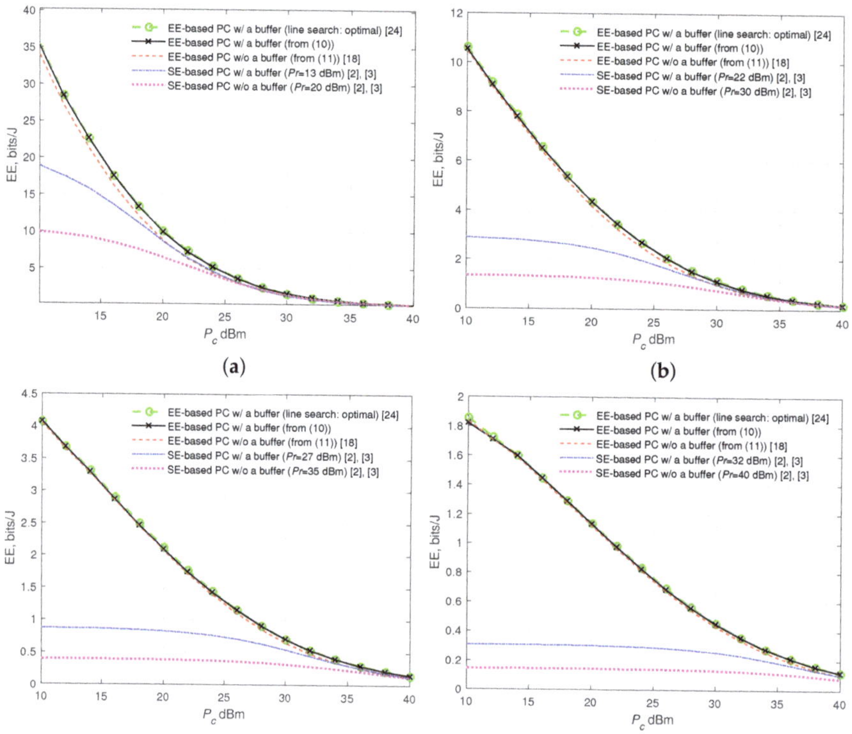

(c) (d)

Figure 8. EE across P_c with EE- and SE-based power control (PC) for RN. (**a**) $d_{rd} = 0.3$ km. (**b**) $d_{rd} = 0.5$ km. (**c**) $d_{rd} = 0.7$ km. (**d**) $d_{rd} = 0.9$ km.

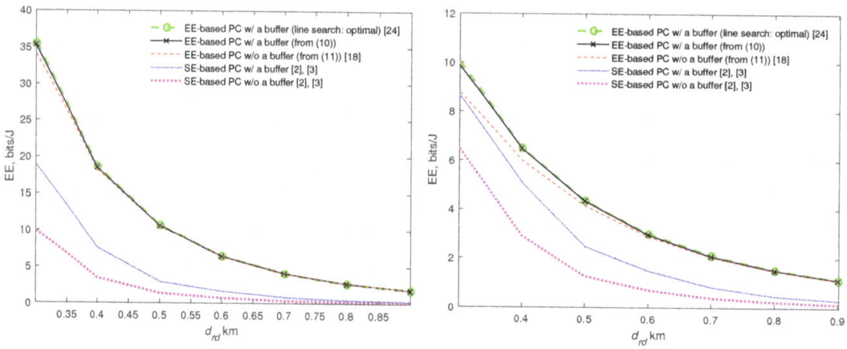

(a) (b)

Figure 9. *Cont.*

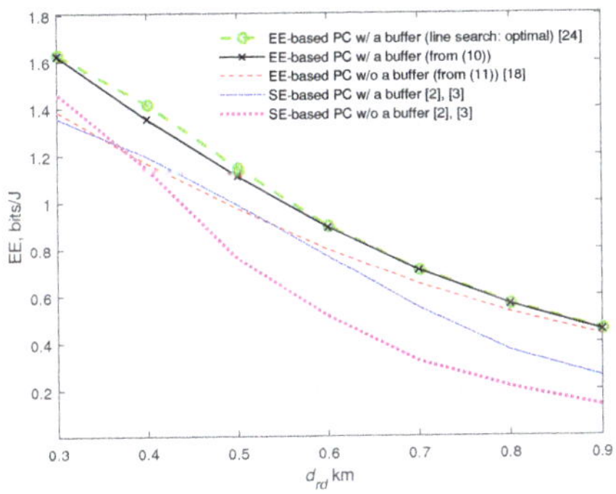

(c)

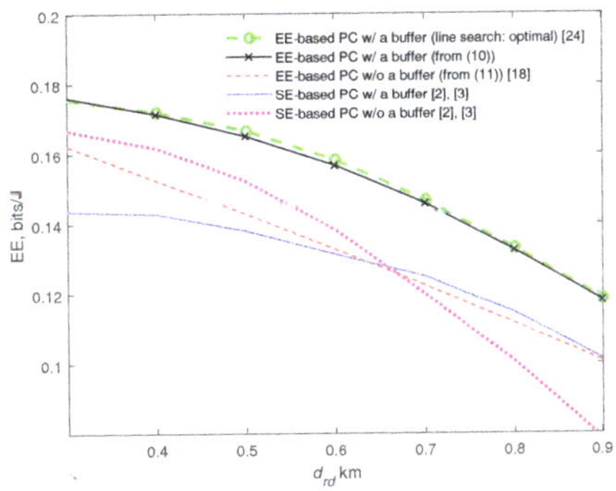

(d)

Figure 9. EE across d_{rd} for RN with a power control (PC) for RN. (**a**) $P_c = 10$ dBm. (**b**) $P_c = 20$ dBm. (**c**) $P_c = 30$ dBm. (**d**) $P_c = 40$ dBm.

7. Conclusions

In this study, a relay system has been considered, in which a ground source node (SN) sends information to a destination node (DN) through an unmanned aerial vehicle relay node (RN) with a buffer. When the transmit power of SN is fixed, we have proposed a two-dimensional surface fitting method to simply obtain the near-optimal transmit power of RN to maximize the energy efficiency (EE) of the RN. To this end, the RN needs its power consumption model and the distance from a DN. From the rigorous numerical results, it has been verified that the proposed power control method can significantly improve the EE of RN compared to a fixed power control method based on spectral efficiency, and achieve near-optimal performance.

Author Contributions: Conceptualization, J.J. and H.L.L.; methodology, J.J.; software, J.Z.; validation, J.Z and X.K.; investigation, J.Z.; writing, original draft preparation, J.J.; review and editing, H.L.L, J.Z. and X.K.; visualization, J.J. and H.L.L; supervision, J.J.; project administration, J.J.; funding acquisition, J.J.

Funding: This research was supported by the National Research Foundation of Korea (NRF) grant (2018R1A4A1023826) funded by the Korean government (MSIT).

Conflicts of Interest: The authors declare no conflicts of interest.

References

1. Cover, T.M.; El Gammal, A. Capacity theorems for relay channels. *IEEE Trans. Inf. Theory* **1979**, *25*, 572–584. [CrossRef]
2. Rankov, B.; Wittneben, A. Spectral efficient protocols for half-duplex fading relay channels. *IEEE J. Sel. Areas Commun.* **2007**, *25*, 379–389. [CrossRef]
3. Zhao, J.; Kuhn, M.; Wittneben, A.; Bauch, G. Cooperative transmission schemes for decode-and-forward relaying. In Proceedings of the 2007 IEEE 18th International Symposium on Personal, Indoor and Mobile Radio Communications, Athens, Greece, 3–7 September 2007; pp. 1–5.
4. Darsena, D.; Gelli, G.; Iudice, I.; Verde, F. Widely-linear transceiver design for amplify-and-forward MIMO relaying. In Proceedings of the 2016 IEEE Sensor Array and Multichannel Signal Processing Workshop (SAM), Rio de Janeiro, Brazil, 10–13 July 2016; pp. 1–5.
5. Barman Roy, S.; Madhukumar, A.S.; Joung, J. On joint Pareto frontier in multiple access and relay rate regions with Rayleigh fading. *IEEE Trans. Veh. Technol.* **2017**, *66*, 3777–3786.
6. Joung, J.; Sayed, A.H. Multiuser two-way amplify-and-forward relay processing and power control methods for beamforming systems. *IEEE Trans. Signal Process.* **2010**, *58*, 1833–1846. [CrossRef]
7. Joung, J.; Sayed, A.H. User selection methods for multiuser two-way relay communications using space division multiple access. *IEEE Trans. Wirel. Commun.* **2010**, *9*, 2130–2136. [CrossRef]

8. Patel, C.S.; Stuber, G.L. Channel estimation for amplify and forward relay based cooperation diversity systems. *IEEE Trans. Wirel. Commun.* **2007**, *6*, 2348–2356. [CrossRef]
9. Unger, T.; Klein, A. Applying relay stations with multiple antennas in the one- and two-way relay channel. In Proceedings of the IEEE 18th International Symposium on Personal, Indoor and Mobile Radio Communications, (PIMRC), Athens, Greece, 3–7 September 2007.
10. Chun, B.; Jeong, E.R.; Joung, J.; Oh, Y.; Lee, Y.H. Pre-nulling for self-interference suppression in full-duplex relays. In Proceedings of the 2009 APSIPA Annual Summit and Conference, Sapporo, Japan, 4–7 October 2009; pp. 949–952.
11. Joung, J.; Sun, S. Design and performance evaluation of multiple AF-relay processing in multi-cell environment. In Proceedings of the 2011 IEEE 73rd Vehicular Technology Conference (VTC Spring), Budapest, Hungary, 15–18 May 2011; pp. 1–5.
12. Joung, J.; Sun, S. A simple network-power-saving resource allocation method for OFDMA cellular networks with multiple relays. In Proceedings of the 2011 IEEE International Conference on Acoustics, Speech and Signal Processing (ICASSP), Prague, Czech Republic, 22–27 May 2011; pp. 2504–2507.
13. Hammerström, I.; Kuhn, M.; Eşli, C.; Zhao, J.; Wittneben, A.; Bauch, G. MIMO two-way relaying with transmit CSI at the relay. In Proceedings of the IEEE Signal Processing Advances in Wireless Communications (SPAWC), Helsinki, Finland, 17–20 June 2007.
14. Ye, H.T.; Kang, X.; Liang, Y.C.; Joung, J. Full-duplex wireless-powered IoT networks with unmanned aerial vehicle. In Proceedings of the The 9th International Conf. ICT Convergence (ICTC), Jeju Island, Korea, 17–19 October 2018; pp. 1–4.
15. Zhang, S.; Zhang, H.; He, Q.; Bian, K.; Song, L. Joint trajectory and power optimization for UAV relay networks. *IEEE Commun. Lett.* **2018**, *22*, 161–164. [CrossRef]
16. Ye, H.T.; Kang, X.; Liang, Y.C.; Joung, J. Optimal time allocation for full-duplex wireless-powered IoT networks with unmanned aerial vehicle. In Proceedings of the IEEE International Conference on Communications (ICC), Shanghai, China, 20–24 May 2019; pp. 1–6.
17. Kang, H.; Joung, J.; Ahn, J.; Kang, J. Secrecy-aware altitude optimization for quasistatic UAV base station without eavesdropper location information. *IEEE Commun. Lett.* **2019**, *23*, 851–854. [CrossRef]
18. Joung, J.; Sun, S. Power efficient resource allocation for downlink OFDMA relay cellular networks. *IEEE Trans. Signal Process.* **2012**, *60*, 2447–2459. [CrossRef]
19. Singh, K.; Ku, M.L.; Biswas, S.; Ratnarajah, T. Energy-efficient subcarrier pairing and power allocation for DF relay networks with an eavesdropper. *Energies* **2017**, *10*, 1953. [CrossRef]
20. Zhang, J.; Zeng, Y.; Zhang, R. Spectrum and energy efficiency maximization in UAV-enabled mobile relaying. In Proceedings of the IEEE International Conference on Communications (ICC), Paris, France, 21–25 May 2017; pp. 1–6.
21. Xia, B.; Fan, Y.; Thompson, J.; Poor, H.V. Buffering in a three-node relay network. *IEEE Trans. Wirel. Commun.* **2008**, *7*, 4492–4496. [CrossRef]
22. Zlatanov, N.; Schober, R. Buffer-aided relaying with adaptive link selection-fixed and mixed rate transmission. *IEEE Trans. Inf. Theory* **2013**, *59*, 2816–28404. [CrossRef]
23. Qiao, D. Effective capacity of buffer-aided full-duplex relay systems with selection relaying. *IEEE Trans. Veh. Technol.* **2016**, *64*, 117–129. [CrossRef]
24. Hajipour, J.; Niya, J.M.; Ng, D.W.K. Energy-efficient resource allocation in buffer-aided wireless relay networks. *IEEE Trans. Wirel. Commun.* **2017**, *16*, 6648–6659. [CrossRef]
25. Nakai, R.; Oiwa, M.; Sugiura, S. Generalized buffer-state-based relay selection for fixed-rate buffer-aided cooperative systems. In Proceedings of the IEEE 85th Vehicular Technology Conference (VTC Spring), Sydney, NSW, Australia, 4–7 June 2017; pp. 1–5.
26. Haneda, K.; Kahra, E.; Wyne, S.; Icheln, C.; Vainikainen, P. Measurement of loop-back interference channels for outdoor-to-indoor full-duplex radio relays. In Proceedings of the Fourth European Conference Antennas and Propagation, Barcelona, Spain, 12–16 April 2010; pp. 1–5.
27. Cumanan, K.; Ding, Z.; Rahulamathavan, Y.; Molu, M.M.; Chen, H. Robust MMSE beamforming for multiantenna relay networks. *IEEE Trans. Veh. Technol.* **2017**, *66*, 3900–3912. [CrossRef]
28. Joung, J.; Sun, S. Phase rotation and link selection methods for DSTTD-based two-path relay systems. *IEEE Commun. Lett.* **2011**, *15*, 278–280. [CrossRef]

29. Joung, J. Beamforming vector design for regenerative wired two-way relay systems. *IET Electron. Lett.* **2017**, *53*, 596–598. [CrossRef]
30. Joung, J.; Choi, J. Linear precoder design for an AF two-way MIMO relay node with no source node precoding. *IEEE Trans. Veh. Technol.* **2017**, *66*, 10526–10531. [CrossRef]
31. LTE; E-UTRA. RF Requirements for LTE Pico Node B; ETSI std., 136 931 V9.0.0; 2011. Available online: https://www.etsi.org/deliver/etsi_tr/136900_136999/136931/09.00.00_60/tr_136931v090000p.pdf (accessed on 22 August 2019).
32. Joung, J.; Sun, S. Energy efficient power control for distributed transmitters with ZF-based multiuser MIMO precoding. *IEEE Commun. Lett.* **2013**, *17*, 1766–1769. [CrossRef]
33. Joung, J.; Chia, Y.K.; Sun, S. Energy-efficient, large-scale distributed-antenna system (L-DAS) for multiple users. *IEEE J. Sel. Topics Signal Process.* **2014**, *8*, 954–965. [CrossRef]
34. Joung, J.; Ho, C.K.; Sun, S. Spectral efficiency and energy efficiency of OFDM systems: Impact of power amplifiers and countermeasures. *IEEE J. Sel. Areas Commun.* **2014**, *32*, 208–220, doi:10.1109/JSAC.2014.141203. [CrossRef]
35. Rappaport, T.S. *Wireless Communications*, 2nd ed.; Prentice Hall: Upper Saddle River, NJ, USA, 2002.
36. Joung, J.; Ho, C.K.; Sun, S. Power amplifier switching (PAS) for energy efficient systems. *IEEE Wirel. Commun. Lett.* **2013**, *2*, 14–17. [CrossRef]
37. Joung, J.; Sun, S. EMA: Energy-efficiency-aware multiple access. *IEEE Commun. Lett.* **2014**, *18*, 1071–1074. [CrossRef]
38. Joung, J.; Ho, C.K.; Adachi, K.; Sun, S. A survey on power-amplifier-centric techniques for spectrum and energy efficient wireless communications. *IEEE Commun. Surv. Tutor.* **2014**, *17*, 315–333. [CrossRef]
39. Zeng, Y.; Xu, J.; Zhang, R. Energy minimization for wireless communication with rotary-wing UAV. *IEEE Trans. Wirel. Commun.* **2019**, *18*, 2329–2345. [CrossRef]
40. Antoniou, A.; Lu, W.S. *Practical Optimization: Algorithms and Engineering Applications*, 1st ed.; Springer: New York, NY, USA, 2007.
41. Joung, J. Machine learning-based antenna selection in wireless communications. *IEEE Commun. Lett.* **2016**, *20*, 2241–2244. [CrossRef]
42. Jadoon, M.A.; Kim, S. Relay selection algorithm for wireless cooperative networks: A learning-based approach. *IET Commun.* **2017**, *11*, 1061–1066. [CrossRef]
43. Abdelreheem, A.; Omer, O.A.; Esmaiel, H.; Mohamed, U.S. Deep learning-based relay selection in D2D millimeter wave communications. In Proceedings of the 2019 International Conference on Computer and Information Sciences (ICCIS), Milan, Italy, 30 October–1 November 2019; pp. 1–5.
44. Chapra, S.C. *Applied Numerical Methods with MATLAB for Engineerrs and Scientists*, 3rd ed.; Mc Graw Hill: Singapore, 2012.

energies

MDPI

Article

Energy-Neutral Operation Based on Simultaneous Wireless Information and Power Transfer for Wireless Powered Sensor Networks

Hyun-Ho Choi [1] **and Jung-Ryun Lee** [2,*]

1 Department of Electrical, Electronic and Control Engineering, Hankyong National University, Anseong 17579, Korea; hhchoi@hknu.ac.kr
2 School of the Electrical Engineering, Chung-Ang University, Seoul 06974, Korea
* Correspondence: jrlee@cau.ac.kr

Received: 21 August 2019; Accepted: 04 October 2019; Published: 10 October 2019

Abstract: For energy-neutral operation (ENO) of wireless sensor networks (WSNs), we apply a wireless powered communication network (WPCN) to a WSN with a hierarchical structure. In this hierarchical wireless powered sensor network (WPSN), sensor nodes with high harvesting energies and good link budgets have energy remaining after sending their data to the cluster head (CH), whereas the CH suffers from energy scarcity. Thus, we apply the simultaneous wireless information and power transfer (SWIPT) technique to the considered WPSN so that the sensor nodes can transfer their remaining energy to the CH while transmitting data in a cooperative manner. To maximize the achievable rate of sensing data while guaranteeing ENO, we propose a novel ENO framework, which provides a frame structure for SWIPT operation, rate improvement subject to ENO, SWIPT ratio optimization, as well as clustering and CH selection algorithm. The results of extensive simulations demonstrate that the proposed ENO based on SWIPT significantly improves the achievable rate and reduces the energy dissipated in the network while guaranteeing ENO, in comparison with the conventional schemes without SWIPT.

Keywords: energy-neutral operation; wireless powered sensor network; simultaneous wireless information and power transfer; energy harvesting; clustering

1. Introduction

With the emergence of energy harvesting (EH) techniques, sensors can be equipped with EH modules to acquire additional energy from the ambient resources (i.e., solar radiation, wind, vibrations, radio-frequency (RF) power, etc.) Such EH sensors do not break down due to energy shortages as long as the energy consumption is less than the harvested energy, so they can operate perpetually with a desired performance level, which is called energy neutral operation (ENO) [1].

According to the controllability of energy sources, ENO approaches can be classified as ambient energy harvesting (AEH) or wireless power transfer (WPT). AEH is the process of transforming any ambient resource, such as solar radiation or wind, into readily utilizable energy [2]. In AEH, it is difficult to control the amount of energy supplied due to the random nature of the employed energy sources. Thus, by predicting the energy source activity, various adaptive energy management schemes, which control the duty cycles [3], transmission power [4], sampling rates [5], and routing paths [6] of sensor nodes, have been proposed to guarantee ENO. Meanwhile, WPT uses a controllable RF power source, such as a power beacon and hybrid access point (HAP) (While a power beacon only acts as a power transmitter, a HAP acts as both a power transmitter and a communication gateway.) Various radio resources (e.g., time, bandwidth, waveform, antennas, etc.) can be controlled to transfer the RF energy efficiently while ensuring ENO [7,8]. Such controllability at the RF power source adds a new dimension to the system optimization, and various optimization problems have been addressed to maximize system performance while ensuring ENO [9,10].

The potential of WPT has recently begun to emerge in two major applications: wireless powered communication networks (WPCNs) [11] and simultaneous wireless information and power transfer (SWIPT) [12]. A WPCN consists of a dedicated power source (e.g., HAP) and wireless devices (e.g., sensors), where the wireless devices are powered by the RF waves sent from the HAP and then transmit data to the HAP using the harvested energy. On the other hand, SWIPT is a technique that enables both wireless information transmission (WIT) and wireless energy transfer (WET) to be attained simultaneously via the same electromagnetic wave [13]. To this end, one of two mode-switching techniques, namely power splitting (PS) or time switching (TS), is used to balance the ratio of WIT to WET [14].

One of the main challenges in the operation of wireless sensor networks (WSNs) is the limited battery time of the sensor nodes. WSNs typically consist of massive numbers of sensor nodes, so it is costly and impractical to replace their batteries regularly [15]. It is also very serious that sensing errors and link failures often occur when the battery life of a sensor node is almost over. Therefore, extending the lifetimes of sensor nodes while maintaining their sensing performance is a major problem in WSNs [16,17].

To address this issue, we apply the WPT technology to a WSN. Firstly, we apply the basic WPCN concept to a hierarchical network structure, as many WSN applications use this structure to reduce the total cost of the transmission links [18]. In this wireless powered sensor network (WPSN) with a hierarchical structure, all of the nodes in the cell harvest wireless energy from the HAP, as in a WPCN. However unlike in a WPCN, the sensor nodes are clustered and transmit the sensing data to the cluster head (CH) without directly transmitting the data to the HAP. The CH gathers and aggregates all of the sensing data received from its member nodes and then transmits the aggregated data to the HAP at one time.

If all of the sensor nodes perform the same task and thus generate sensing data of the same size, the maximum rate of the sensing data collected in the WSN will be limited by the worst sensor node with low harvesting energy and poor link budget [19,20]. Thus, all of the sensor nodes only need to support the same data rate as the worst node so that some sensor nodes may have energy remaining after transmitting their sensing data to the CH. On the other hand, the CH needs more processing for the reception and aggregation of multiple sensing data and has to transmit the aggregated data to the HAP via an uplink; thus, it requires more energy in general (i.e., the CH becomes the highest energy-consuming node in the cluster with high probability). Considering this situation, we apply the SWIPT technique to the considered WPSN so that the sensor nodes could transfer their remaining energy to the CH while transmitting data in a cooperative way. This approach can increase the sensing data rate in the cluster while guaranteeing the ENO of sensor nodes because the CH receives additional energy from its member nodes and the sensor nodes give up only the remaining energy. The objective of our study is to maximize the achievable rate of sensing data while guaranteeing ENO in the considered WPSN. The main contributions can be summarized as follows:

- We design a frame structure to operate SWIPT in the hierarchical WPSN structure. The frame is divided into WET, SWIPT, and WIT slots, and each sensor node uses either a PS or TS method in the allocated SWIPT slot.
- We numerically express the achievable rate of sensing data collected in each cluster subject to the guarantee of ENO of the sensor nodes and obtain the total energy dissipated in the cluster depending on the use of SWIPT.
- We develop an algorithm that finds the optimal SWIPT ratio in terms of PS and TS to maximize the achievable rate of sensing data in the cluster.
- Finally, we design a clustering and CH selection algorithm based on the K-means clustering algorithm to maximize the achievable rate in the entire network.

The rest of this paper is organized as follows. In Section 2, we survey related previous studies and explain the originality of our study. In Section 3, we describe the considered WPSN system and introduce our basic approach. In Section 4, we explain the proposed ENO framework in terms of frame structure, the details of ENO, the optimal SWIPT ratio, and clustering and CH selection. In Section 5, the simulation results are discussed. Finally, we present the conclusions in Section 6.

2. Related Works

In this section, we survey the previous studies on WPT-based ENO, which are directly related to the proposed approach.

2.1. WPCN

WPCN basically guarantees ENO because all of the devices in the cell use only the energy supplied by the HAP or power beacon. Thus, most studies on WPCNs have been focused on system optimization considering resource allocation [21], beamforming [22], cooperative communication [23], and full-duplex communication [24]. These topics have been investigated in a flat network structure where the wireless device directly transmits data to the HAP, while hierarchical networks like WSNs have not been considered.

A few studies have addressed hierarchical structures in WPCNs. For instance, cluster-based cooperation in a WPCN was proposed in [25], where an HAP with multiple antennas exploited energy beamforming to focus more transferred power to the CH considering the performance to be limited by the high energy consumption of the CH. Then, joint optimization of the energy beamforming design, transmit times of all of the nodes, and transmission power of the CH was performed to maximize the minimum data rate achievable among all of the nodes (i.e., max-min throughput). In [26], a WPSN consisting of one HAP, a near cluster, and the corresponding far cluster was considered and a cluster cooperation concept was proposed. If the sensors in the near cluster do not have their own information to transmit, acting as relays, they can help the sensors in the far cluster forward information to the HAP in an amplify-and-forward manner. In [27], a WPSN was divided into several layers and the exact border of each layer was obtained in order to alleviate the doubly near-far effect, and the energy broadcasted by the HAP and radius of each layer were optimized jointly. Furthermore, a multi-hop WPCN based on user-cooperative multi-hop transmission was considered in [28] to improve the throughput fairness, and the max-min throughput was optimized by resource allocation (i.e., transmission power and time). In a WPCN environment, these previous studies considered the WET from the HAP without using SWIPT technology.

2.2. SWIPT in WSN

SWIPT has been applied to WSNs in various ways to overcome their energy limitations. Numerous surveys of SWIPT in WSNs have been published recently [29–31]. For instance, [29] summarized the current research on SWIPT-based cooperative sensor networks, in which SWIPT is applied to WSNs in terms of dual-hop and multi-hop relays. Meanwhile, [30] focused on the integral aspects of SWIPT in other prominent networks, such as device-to-device networks, vehicular ad hoc networks, wireless body area networks, WSNs, and so on, and presented open issues and challenges in SWIPT application. In [31], an overview of SWIPT/WPT-enabled WSNs was provided and future directions were suggested.

Meanwhile, [32] described a WSN consisting of multiple clusters and a sink node, in which the CH of each cluster performs SWIPT to give energy to relay nodes with low energies. In [33], a SWIPT-powered sensor network was considered in which each source node operates SWIPT as both an information transmitter and an energy transferrer, and the destination node works as an information receiver while the other nodes work as energy harvesters. Furthermore, [34] focused on the deployment of a WSN and its routing strategy when SWIPT is applied to the WSN. The basic idea is to reduce the total recharging cost to enhance the lifespan of the WSN. In addition, [35] proposed a dynamic routing algorithm for a renewable WSN with SWIPT. SWIPT has been adopted in many studies to overcome the energy limitations of WSNs, but SWIPT has not yet been applied in the WPCN environment. Furthermore, these previous studies were focused on increasing the lifetimes of WSNs rather than guaranteeing ENO.

2.3. ENO Based on WPT

ENO based on WPT has been investigated in some recent work. For example, the feasibility of an energy-neutral Internet of Things (IoT) network was studied in [36] by joint utilization of EH and WPT technologies, and design guidelines for energy-neutral wireless powered networks were presented. In [37], duty cycle operation for a self-powering dual mode SWIPT system in which a sensor node adaptively controls single-tone or multi-tone communication mode was introduced and an adaptive mode switching problem was solved to maximize the achievable rate under ENO to realize a battery-free IoT network. Meanwhile, an energy management scheme for a WPSN was proposed in [38], which maximizes the RF energy transfer efficiency while guaranteeing ENO, and the proposed energy management scheme was implemented in an actual testbed to verify its operation. In [39], an adaptive duty cycle control algorithm was proposed to prevent the energy storage of a sensor node from being depleted and its ENO was validated in a multi-antenna WPSN testbed that dynamically steered a microwave beam towards a sensor node. In addition, the energy neutral Internet of drones was conceptualized in [40] to overcome the energy limitations for continuous operation of drones and to enhance the connectivity between drones. Communication and networking architectures and protocols for drones energized by WPT were presented. Although ENO was considered in SWIPT-enabled networks or WPCNs in these works, to the best of our knowledge, ENO considering both SWIPT and WPCNs has not been investigated yet.

3. System Description

Figure 1 illustrates the considered WPSN, which includes one HAP and a number of sensor nodes in a cell. The HAP is a power source that wirelessly transfers energy to all of the sensor nodes and also acts as a sink node that collects sensing information in the cell. Only the HAP has a stable energy supply, but the sensors do not have any embedded energy source or battery. The HAP periodically broadcasts an RF wave, and all of the nodes in the cell harvest energy from this RF signal. Considering the hierarchical structure, the sensor nodes are grouped into a number of clusters and one CH is selected from among the member nodes in each cluster. Thus, all of the sensor nodes send sensing data to their CH instead of directly to the HAP. The CH aggregates all of the sensing information received from its member sensor nodes as well as its own sensing information (i.e., data fusion), then sends the aggregated data to the HAP at once.

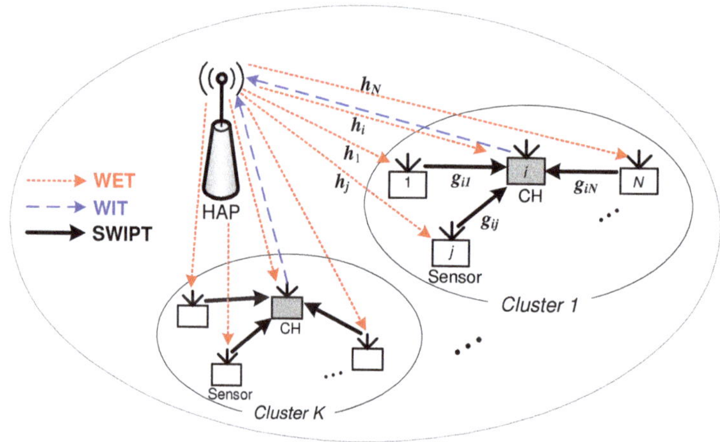

Figure 1. Considered wireless powered sensor network.

Assuming that all of the sensors perform the same sensing task, such as temperature, humidity, or fire sensing, the bit size of the sensing data transmitted by each sensor node is the same. However, the sensor nodes inherently harvest different amounts of energy depending on their distances from the HAP and consume different amounts of energy for transmission depending on their distances from the CH. Incidentally, the maximum rate of sensing data collection in the WSN is limited by the worst sensor node, which has a low harvesting energy and poor link budget [19,20]. Therefore, all of the sensor nodes only need to support the same data rate as the worst node so that some sensor nodes may have energy remaining after transmitting their sensing data to the CH. Since the CH usually requires more energy for multiple receptions, processing, and uplink transmission to the HAP, the sensor nodes transfer the remaining energy to the CH while sending their data by using the SWIPT technique, and the CH acquires additional energy while receiving data from its member nodes.

4. Proposed Energy Neutral Operation

In this section, we propose new protocols and algorithms related to the frame structure, ENO, SWIPT, clustering, and CH selection, in order to maximize the achievable rate of sensing data while guaranteeing ENO in the considered WPSN.

4.1. Frame Structure

Figure 2 shows the frame structure for the proposed ENO based on SWIPT. The frame is based on time division multiple access with time division duplexing (TDMA/TDD) and scheduling-based resource allocation is used for conflict-free transmission. Firstly, a beacon signal is broadcasted at the beginning of the frame for frame synchronization and to provide the frame configuration and scheduling information to all of the nodes in the cell. In compliance with the harvest-then-transmit protocol [21], the HAP transmits RF energy during the WET slot with length T_e. During this WET period, all of the sensor nodes harvest energy to be used in the current frame.

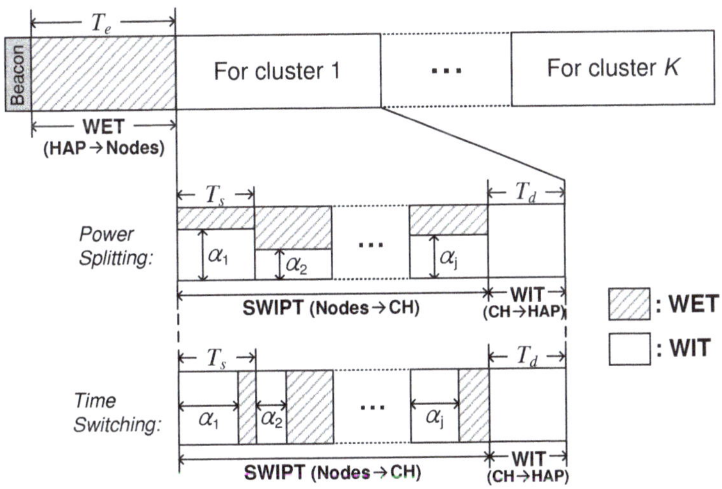

Figure 2. Frame structure for the proposed ENO based on SWIPT.

The remainder of the frame is divided and used for each cluster. The resources allocated to each cluster consist of multiple SWIPT slots for the member nodes in the cluster and one WIT slot for the CH. The total number of SWIPT slots for a certain cluster is the same as the number of member nodes in that cluster, and the length of a SWIPT slot is set to T_S. Each SWIPT slot is reserved for each sensor node through a pre-scheduling mechanism so that access collision does not occur [41]. On the other hand, the last WIT slot is dedicated to the CH only for WIT and has a length of T_d. Hence, the total length of the frame depends on the number of clusters and the number of member nodes in each cluster.

The sensor node transmits the sensing data to its CH in the SWIPT slot allocated to itself. Simultaneously, it transfers the remaining energy (if it exists) to the CH by using SWIPT based on either the PS or TS method. Each sensor node j uses a portion α_j of power or time for WIT and the remaining portion $(1-\alpha_j)$ for WET. During these SWIPT slots, the CH sequentially receives the sensing data while receiving additional energy from its member nodes. Then, the CH aggregates all of the sensing information and transmits the aggregated sensing data to the HAP at the last WIT slot. Here, the CH uses not only the energy initially provided by the HAP but also the energy additionally received from its member nodes. It is noted that we do not consider the ambient energy harvesting between any nodes [42] because the amount of energy harvested is negligible due to the long distance between nodes and the sensor nodes may inherently stay in sleep state except for its transmit and receive periods.

4.2. Energy Neutral Operation

We explain the proposed ENO based on SWIPT in detail in this section. The proposed ENO is performed on a per-cluster basis. Without loss of generality, we can consider only one cluster assuming that the clustering and CH selection are completed in advance (The clustering and CH selection algorithm is presented in Section 4.4.). Suppose that there are N sensor nodes in the cluster. Let $\mathcal{N}$ be the set of sensor nodes in the cluster, i.e., $\mathcal{N} = \{1, 2, \cdots, N\}$. We denote the selected CH node as i, where $\exists i \in \mathcal{N}$, and denote the other sensor nodes as j, where $\forall j \in \mathcal{N} \setminus \{i\}$. For simplicity, we assume that there is channel reciprocity between any two nodes. As illustrated in Figure 1, h_j denotes the channel power gain between the HAP and any node j, and g_{ij} denotes the channel power gain from sensor node j to CH i.

The energy harvested by node j from HAP is expressed as

$$E_j^{\text{har}} = \zeta_j h_j P T_e, \ \forall j \in \mathcal{N} \tag{1}$$

where $0 < \zeta_j < 1$ is an EH efficiency of node j, and P is the constant transmission power of the HAP. The transmission power of sensor node j for transmitting data to the CH during the SWIPT slot is given by

$$P_j = \frac{\eta_j E_j^{\text{har}}}{T_s} = \frac{\eta_j \zeta_j h_j P T_e}{T_s}, \ \forall j \in \mathcal{N} \setminus \{i\} \tag{2}$$

where $0 < \eta_j < 1$ is the ratio of the energy used for only transmission except reception, processing, and circuit operations to the total energy harvested in node j. By Shannon's theorem, the achievable rate of transmission from sensor node j to the CH i is expressed as

$$R_{ij} = T_s \log_2 \left(1 + \frac{g_{ij} P_j}{\sigma^2}\right), \ \forall j \in \mathcal{N} \setminus \{i\} \ \text{(bits/Hz)} \tag{3}$$

where σ^2 is the noise power at the receiving side, and the unit of this rate has bits/Hz reflecting the transmission time T_s.

If SWIPT is not used, as in the conventional method, the CH uses only the energy harvested from the HAP. Thus, the transmission power of CH i for transmitting data to the HAP during the WIT slot is given by

$$P_i = \frac{\eta_i E_i^{\text{har}}}{T_d} = \frac{\eta_i \zeta_i h_i P T_e}{T_d}. \tag{4}$$

Then, the achievable rate of transmission from the CH i to the HAP is given by

$$R_i = T_d \log_2 \left(1 + \frac{h_i P_i}{\sigma^2}\right) \text{ (bits/Hz)}. \tag{5}$$

The maximum rate of sensing data that can be collected in this cluster is limited by the minimum rate of the transmission links [19,20]. Therefore, when SWIPT is not used, the achievable rate of sensing data in this cluster is determined by

$$R^{\text{noSwipt}} = \min \left[\min_{j \in \mathcal{N} \setminus \{i\}} \{R_{ij}\}, R_i \right]. \tag{6}$$

Because the achievable rate of sensing data is determined by the minimum rate of the transmission links, the sensor nodes with higher R_{ij} can consume only the energy required to satisfy this minimum rate and then transmit the remaining energy to the CH using the SWIPT technique. When PS-based SWIPT is applied, node j uses power $\alpha_j P_j$ for WIT and the remaining power $(1 - \alpha_j)P_j$ for WET. Thus, the achievable rate of transmission from sensor node j to CH i is expressed as

$$\begin{aligned} R_{ij}^{\text{PS}} &= T_s \log_2 \left(1 + \tfrac{g_{ij}\alpha_j P_j}{\sigma^2}\right) \\ &= T_s \log_2 \left(1 + \tfrac{\eta_j \zeta_j g_{ij} h_j \alpha_j P T_e}{\sigma^2 T_s}\right), \ \forall j \in \mathcal{N} \setminus \{i\} \text{ (bits/Hz)}. \end{aligned} \tag{7}$$

On the other hand, when TS-based SWIPT is applied, node j uses time $\alpha_j T_s$ for WIT and the remaining time $(1 - \alpha_j)T_s$ for WET. Hence, the achievable rate of transmission from sensor node j to CH i is expressed as

$$\begin{aligned} R_{ij}^{\text{TS}} &= \alpha_j T_s \log_2 \left(1 + \tfrac{g_{ij} P_j}{\sigma^2}\right) \\ &= \alpha_j T_s \log_2 \left(1 + \tfrac{\eta_j \zeta_j g_{ij} h_j P T_e}{\sigma^2 T_s}\right), \ \forall j \in \mathcal{N} \setminus \{i\} \text{ (bits/Hz)}. \end{aligned} \tag{8}$$

In both the PS and TS methods, the amount of energy that CH i additionally obtains is given by

$$E_i^{\text{add}} = \sum_{j \in \mathcal{N} \setminus \{i\}} \zeta_i g_{ij} (1 - \alpha_j) P_j T_s. \tag{9}$$

Then, when SWIPT is used, the transmission power of CH i is updated as

$$P_i^{\text{swipt}} = \frac{\eta_i (E_i^{\text{har}} + E_i^{\text{add}})}{T_d}. \tag{10}$$

Moreover, the achievable rate of transmission from CH i to the HAP is represented as

$$\begin{aligned} R_i^{\text{swipt}} &= T_d \log_2 \left(1 + \frac{h_i P_i^{\text{swipt}}}{\sigma^2}\right) \\ &= T_d \log_2 \left\{1 + \frac{\eta_i h_i (E_i^{\text{har}} + E_i^{\text{add}})}{\sigma^2 T_d}\right\} \\ &= T_d \log_2 \left\{1 + \frac{\eta_i \zeta_i h_i P T_e \left(h_i + \sum_{j \in \mathcal{N} \setminus \{i\}} \eta_j \zeta_j (1 - \alpha_j) g_{ij} h_j\right)}{\sigma^2 T_d}\right\} \text{ (bits/Hz)}. \end{aligned} \tag{11}$$

Finally, the achievable rates of sensing data when SWIPT is used according to PS and TS are respectively expressed as

$$R^{\text{swipt}} = \begin{cases} \min \left[\min_{j \in \mathcal{N} \setminus \{i\}} \left\{ R_{ij}^{\text{PS}} \right\}, R_i^{\text{swipt}} \right], & \text{if PS is used} \\ \min \left[\min_{j \in \mathcal{N} \setminus \{i\}} \left\{ R_{ij}^{\text{TS}} \right\}, R_i^{\text{swipt}} \right], & \text{if TS is used.} \end{cases} \tag{12}$$

In WSNs, energy dissipation can occur because the sensor node does not need to consume more energy than the energy required to attain the achievable rate of sensing data. This metric is important for measuring the energy efficiency in WSNs [43]. The amount of energy dissipated in the sensor node is defined as the difference between the energy harvested and the energy required to satisfy the achievable rate. In the case without using SWIPT, all of the nodes only need to satisfy R^{noSwipt} given as Equation (6). Thus, the amounts of energy dissipated in sensor node j and CH i are respectively calculated as

$$\mathcal{E}_j^{\text{noSwipt}} = \left\{ P_j - \frac{\sigma^2}{g_{ij}} \left(2^{\frac{R^{\text{noSwipt}}}{T_s}} - 1 \right) \right\} T_s, \ \forall j \in \mathcal{N} \setminus \{i\} \tag{13}$$

$$\mathcal{E}_i^{\text{noSwipt}} = \left\{ P_i - \frac{\sigma^2}{h_i} \left(2^{\frac{R^{\text{noSwipt}}}{T_d}} - 1 \right) \right\} T_d. \tag{14}$$

Therefore, the total amount of energy dissipated in the cluster is expressed as

$$\mathcal{E}^{\text{noSwipt}} = \mathcal{E}_i^{\text{noSwipt}} + \sum_{j \in \mathcal{N} \setminus \{i\}} \mathcal{E}_j^{\text{noSwipt}}. \tag{15}$$

On the other hand, when SWIPT is used, all of the nodes only need to satisfy R^{PS} or R^{TS} depending on the SWIPT method used. However, SWIPT is not necessary if $R_i \geq \min_j \{R_{ij}\}$ because the rate of the CH is sufficiently high that the CH does not need to receive energy from its member nodes. In this case, the energy dissipated in the cluster is the same as Equation (15). However, in the opposite case, i.e., if $R_i < \min_j \{R_{ij}\}$, the energy dissipated in sensor node j is zero because it transfers all of the remaining energy to the CH. Thus, the energy dissipated in sensor node j is given by

$$\mathcal{E}_j^{\text{swipt}} = \begin{cases} 0, & \text{if } R_i < \min_j \{R_{ij}\} \\ \mathcal{E}_j^{\text{noSwipt}}, & \text{otherwise.} \end{cases} \tag{16}$$

Moreover, the energy dissipated in CH i is obtained as

$$\mathcal{E}_i^{\text{swipt}} = \left\{ P_i^{\text{swipt}} - \frac{\sigma^2}{h_i} \left(2^{\frac{R^{\text{swipt}}}{T_d}} - 1 \right) \right\} T_d. \tag{17}$$

Finally, when SWIPT is used, the total amount of energy dissipated in the cluster is expressed as

$$\mathcal{E}^{\text{swipt}} = \mathcal{E}_i^{\text{swipt}} + \sum_{j \in \mathcal{N} \setminus \{i\}} \mathcal{E}_j^{\text{swipt}}. \tag{18}$$

Note that it is always true that $\mathcal{E}_j^{\text{swipt}} \geq 0$ and $\mathcal{E}_i^{\text{swipt}} \geq 0$ since $R^{\text{swipt}} \leq R_i^{\text{swipt}}$, so the ENO of all of the nodes is guaranteed.

4.3. Optimal SWIPT Ratio

For a given CH i, the optimal SWIPT ratio that maximizes R^{swipt} is formulated as

$$\begin{aligned} \vec{\alpha}^* &= \arg\max_{\vec{\alpha}} R^{\text{swipt}} \\ &= \arg\max_{\vec{\alpha}} \min \left[\min_{j \in \mathcal{N} \setminus \{i\}} \left\{ R_{ij}^{\text{swipt}} \right\}, R_i^{\text{swipt}} \right] \end{aligned} \tag{19}$$

where $\vec{\alpha} = \{\alpha_j\}, \forall j \in \mathcal{N} \setminus \{i\}$, and $R_{ij}^{\text{swipt}} = R_{ij}^{\text{PS}}$ or R_{ij}^{TS} according to the SWIPT method used. From Equations (7), (8), and (11), as $\vec{\alpha}$ increases, R_{ij}^{swipt} increases but R_i^{swipt} decreases, and vice versa. Thus, there is a tradeoff between $\min_j \left\{ R_{ij}^{\text{swipt}} \right\}$ and R_i^{swipt} according to $\vec{\alpha}$. In this case, the optimal solution of the given max-min problem is found when $\min_j \left\{R_{ij}^{\mathrm{swipt}}\right\} = R_i^{\mathrm{swipt}}$ [44]. To find the optimal $\vec{\alpha}^*$ that satisfies $\min_j \left\{R_{ij}^{\mathrm{swipt}}\right\} = R_i^{\mathrm{swipt}}$, we design an iterative algorithm, as shown in Algorithm 1. Since SWIPT is performed when $\min_j \{R_{ij}\} > R_i$, we set the initial target rate R_i^{tar} to $\min_j \{R_{ij}\}$ and adjust the target rate to make all of the link rates equal. In every step, α_j is determined to satisfy R_i^{tar} according to the PS or TS method. The computational complexity of the proposed SWIPT algorithm is given by $O(L|\mathcal{N}|)$ where L is the number of iterations for the first loop and $|\mathcal{N}|$ is the number of nodes in cluster.

Algorithm 1: SWIPT algorithm

```
Ensure: i is CH and j ∈ N \ {i}
Initialize R_i^tar ← min_j{R_ij}
repeat
    if PS is used then
        α_j ← (2^(R_i^tar / T_s) − 1) σ²T_s / (η_j ζ_j g_ij h_j P T_e)   from Equation (7)
    else if TS is used then
        α_j ← R_i^tar / (T_s log_2(1 + η_j ζ_j g_ij h_j P T_e / (σ² T_s)))   from Equation (8)
    end if
    Update E_i^add according to Equation (9)
    Update R_i^swipt according to Equation (11)
    R_i^tar ← (R_i^swipt + R_i^tar) / 2
until R_i^swipt ≥ R_i^tar
Return α_j, ∀j ∈ N \ {i}
```

4.4. Clustering and CH Selection

Next, we discuss how to form appropriate clusters and select the optimal CH. The proposed clustering and CH selection algorithm is described in Algorithm 2. When the number of clusters is K, let $\mathcal{N}_k$ be the set of nodes in cluster $k \in \{1, 2, \cdots, K\}$. Without loss of generality, we denote i_k as the CH in cluster k, where $\exists i_k \in \mathcal{N}_k$, and j_k as a sensor node in the cluster k, where $\forall j_k \in \mathcal{N}_k \setminus \{i_k\}$. The proposed algorithm is based on the well-known K-means clustering algorithm [45]. Thus, the initial K clusters are determined using the K-means algorithm, and the initial CH in each cluster is determined to be the node closest to the centroid of each cluster. Consequently, the initial cluster and CH are determined by only the distance among sensor nodes without considering the distance between the sensor node and the HAP. However, this distance directly affects the amount of energy harvested, so it is necessary to modify the basic process of the K-means algorithm to make it suitable for the proposed ENO scheme.

The proposed algorithm mainly consists of two parts, similarly to the K-means algorithm. Firstly, each node reselects the best cluster among the predetermined clusters. After calculating $R_{i_k j}$ and R_{i_k} for each CH i_k, node j selects the cluster that provides the highest $R_{i_k j}$ among the clusters satisfying $R_{i_k j} > R_{i_k}$. This step is included so that SWIPT can be performed without decreasing the rate of the predetermined clusters. Thereafter, the best CH is reselected in each cluster. For all nodes in each cluster $\forall i_k \in \{\mathcal{N}_k\}$, the rate of cluster k, R_k^{swipt}, is calculated using Equation (12). Then, the best CH, i_k^*, is determined as

$$\begin{aligned} i_k^* &= \arg\max_{i_k} R_k^{\mathrm{swipt}} \\ &= \arg\max_{i_k} \min\left[\min_{j_k \in \mathcal{N}_k \setminus \{i_k\}} \left\{R_{i_k j_k}^{\mathrm{swipt}}\right\}, R_{i_k}^{\mathrm{swipt}}\right] \end{aligned} \tag{20}$$

where i_k^* can be found by exhaustive searching with a complexity of $O(|\mathcal{N}_k|)$. This clustering and CH selection algorithm is repeated until all of the cluster sets and CHs are no longer changed. The computational complexity of this algorithm is given by $O(N^2)$ because it is based on k-means algorithm, which has a time complexity of $O(n^2)$ where n is the input data size [45].

Algorithm 2: Clustering and CH selection

Ensure: $\mathcal{N}_k$ is the set of nodes in cluster $k \in \{1, 2, \cdots, K\}$
i_k is the CH in the cluster k, $\exists i_k \in \mathcal{N}_k$
j_k is a sensor node in the cluster k, $\forall j_k \in \mathcal{N}_k \setminus \{i_k\}$
Initialize $\mathcal{N}_k$ and $i_k \leftarrow$ K-means algorithm [45]
repeat
  /* Select the best cluster in the network */
  for *each node* $j \in \{\mathcal{N}_1 \cup \cdots \cup \mathcal{N}_K\}$ **do**
    for *each CH* i_k **do**
      Determine $R_{i_k j}$ from Equation (3)
      Determine R_{i_k} from Equation (5)
    end for
    if $\exists i_k \in \{i_k | R_{i_k j} > R_{i_k}\} \neq \emptyset$ **then**
      j's cluster $\leftarrow \arg_k \max\{R_{i_k j}\}$ for $\exists i_k \in \{i_k | R_{i_k j} > R_{i_k}\}$
    else
      j's cluster $\leftarrow \arg_k \max\{R_{i_k j}\}$ for $\forall i_k$
    end if
  end for
  /* Select the best CH in each cluster */
  for *each cluster* k **do**
    for *each node* $i_k \in \{\mathcal{N}_k\}$ **do**
      Calculate R_k^{swipt} from Equation (12)
    end for
    CH $i_k^* \leftarrow \arg_{i_k} \max\{R_k^{\text{swipt}}\}$ from Equation (20)
  end for
until $\forall \mathcal{N}_k$ *and* $\forall i_k$ *are no longer changed*

5. Results and Discussion

Table 1 summarizes the parameters used in the simulation. We evaluated the performance by adjusting the number of nodes in the network, number of clusters, and network size in appropriate ranges. We consider a square network area in an indoor environment, place the HAP at the center of the area, and randomly distribute the sensor nodes. Assuming all of the sensor nodes to be homogeneous, their EH efficiencies (ζ) are all equal to 0.8 [21]. The ratio of energy used for transmission (η) is set to 0.9 for the sensor nodes and 0.7 for the CH. For ease of exposition, we consider a simple distance-dependent path loss model given by $h_i = Gd_i^{-\gamma}$ and $g_{ij} = Gd_{ij}^{-\gamma}$ assuming the channel fading effect to be averaged out over the frame and all of the channels to be reciprocal [22–24]. Here, d_i is the distance between the HAP and CH i, d_{ij} is the distance between nodes i and j, G refers to the average power attenuation at a reference distance of 1 m and is set to -30 dB, and γ is the path loss exponent, which is set to 2.5 [46]. Moreover, we set the length of the WET slot (T_e) to 5 s and set the lengths of the SWIPT slot (T_s) and WIT slot (T_d) both equal to 0.1 s, assuming the aggregated sensing data to have the same size as the individual sensing data through proper data fusion [47].

Table 1. Parameter Setup.

Parameter	Value
Number of sensor nodes in WPSN	N = 100∼500 (default = 300)
Number of clusters	K = 2∼20 (default = 10)
Width of the square network	W = 10∼50 m (default = 30 m)
Transmission power of HAP	P = 46 dBm
EH efficiency	$\zeta_j = 0.8, \forall j \in \mathcal{N}$
Ratio of energy used for transmission in sensors	$\eta_j = 0.9, \forall j \in \mathcal{N} \setminus \{i\}$
Ratio of energy used for transmission in CH	$\eta_i = 0.7$
Noise spectral density	−174 dBm/Hz
Noise figure	9 dB
Channel bandwidth	1 MHz
Channel power gains	$h_i = Gd_i^{-\gamma}, g_{ij} = Gd_{ij}^{-\gamma}$
Power attenuation at a reference distance of 1 m	$G = -30$ dB
Path loss exponent	$\gamma = 2.5$
Length of WET slot	$T_e = 5$ s
Length of SWIPT slot	$T_s = 0.1$ s
Length of WIT slot	$T_d = 0.1$ s
Number of simulation trials	1000

For performance comparison, we consider the following five schemes:

1. *LEACH*: The CH is chosen randomly based on the stochastic threshold algorithm of the LEACH protocol [41]. The other sensor nodes are connected to the nearest CH and do not use SWIPT.
2. *K-means*: The clusters are created using the K-means clustering algorithm, and the CH is chosen as the node closest to the centroid of each cluster. SWIPT is not used.
3. *Non-SWIPT*: The clusters are created using the K-means clustering algorithm, but the CH is chosen as the node that maximize R^{noSwipt} given by Equation (6). This scheme is the baseline for checking only the effect of the WPSN without SWIPT. The performance of this scheme is given by Equations (6) and (15).
4. *ENO with PS*: The proposed clustering and CH selection algorithm is used, and PS-based SWIPT is applied. The performance is given by Equations (12) and (18).
5. *ENO with TS*: This scheme is the same as ENO with PS, except that TS-based SWIPT is applied.

Figure 3 presents the clustering results of an example case when 300 sensor nodes are deployed randomly and grouped into five clusters in a square network with a width of 30 m. Note that the LEACH protocol is omitted because the CH is randomly selected and continuously changed in LEACH. The size of each node point represents the amount of energy dissipated in the sensor node. The smaller the dot, the less energy is wasted. The results of the K-means algorithm show that the five clusters are geographically well divided. However, there is dissipated energy in most of the nodes because the K-means algorithm does not use SWIPT. Since the non-SWIPT scheme uses the K-means algorithm for clustering, the cluster set is the same, but the selected CH is different. This behavior occurs because the non-SWIPT scheme considers R_i, as shown in Equation (6), and eventually selects the CH that provides a higher rate. In this non-SWIPT case, most of the nodes have the dissipated energy distributions similar to those in the K-means algorithm because SWIPT is not used too. Figure 3c,d show the proposed SWIPT-based ENO schemes. There is no significant difference between the PS and TS methods. It is evident that the clustering results obtained using the proposed schemes are different from those generated using the K-means and non-SWIPT schemes. In addition, there is little energy dissipation at the nodes in the clusters other than Cluster 1 in the lower left corner. This is because each node transfers its remaining energy to the CH through SWIPT and the CH uses it to transmit its data. Meanwhile, SWIPT is not used in Cluster 1 because the rate of the CH is higher than the rates of the other sensor nodes (i.e., $R_i > \min_j\{R_{ij}\}$), so Cluster 1 has a dissipated energy distribution similar to that in the non-SWIPT scheme.

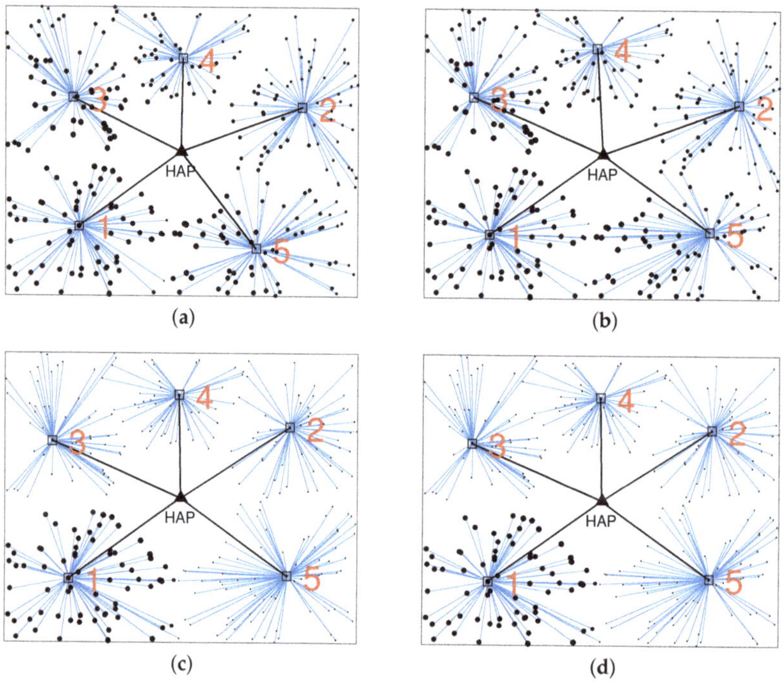

Figure 3. Clustering results of an example case when N = 300, K = 5, and W = 30 m: (**a**) K-means, (**b**) non-SWIPT, (**c**) ENO with PS, and (**d**) ENO with TS.

Figure 4 depicts the achievable rate and energy dissipated in each cluster of Figure 3. In the case of LEACH, both the achievable rate and energy dissipated are the worst because the CH is selected randomly without considering each link rates (i.e., R_{ij} and R_i). In the K-means case, the performance is improved because R_{ij} is considered for clustering and CH selection. In the non-SWIPT case, the performances are better than in the K-means case because R_i is additionally considered for CH selection. In the case of the proposed ENO with PS/TS, the achievable rate is improved more than in the non-SWIPT case because SWIPT is performed while considering both R_{ij} and R_i, and the energy dissipated in each cluster is close to zero except in Cluster 1. This result occurs because each sensor node in the cluster transfers the remaining energy to the CH using SWIPT. This energy is used by the CH to increase the achievable rate of the cluster.

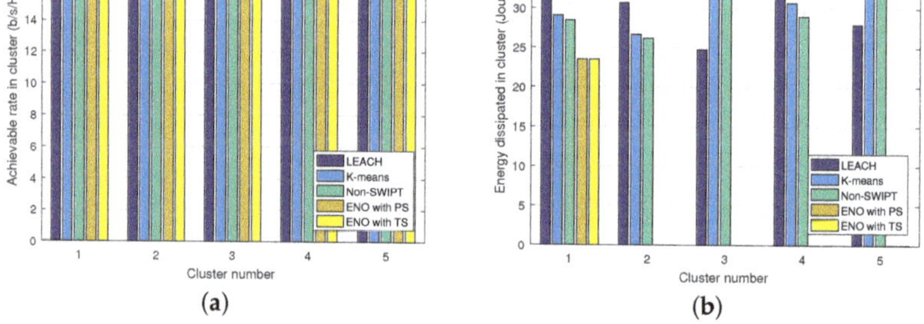

Figure 4. (**a**) Achievable rate and (**b**) energy dissipated in each cluster.

Figure 5 shows the average achievable rate in cluster and the total energy dissipated in network versus the number of clusters (K) when the number of nodes in network (N) is 300 and the width of the square network (W) is 30 m. As K increases, the achievable rate increases in all schemes. These increases occur because the distance between nodes decreases on average as K increases. The LEACH, K-means, and non-SWIPT schemes show better achievable rates in that order. The proposed ENO with PS and ENO with TS have similar performances and outperform the other conventional schemes. Moreover, as K increases, the amount of energy dissipated decreases in all schemes, because the average distance between nodes decreases and the variance of link rates becomes smaller as K increases. The three conventional schemes have similar dissipated energy, which is greater than that in the two proposed ENO schemes.

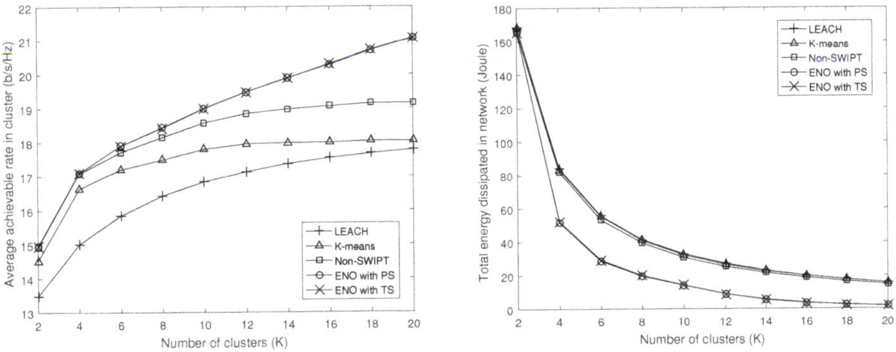

(a) (b)

Figure 5. (**a**) Average achievable rate in cluster and (**b**) total energy dissipated in network vs. number of clusters (K) when N = 300 and W = 30 m.

Figure 6 shows the average achievable rate in cluster and the total energy dissipated in network versus the number of nodes in network (N) when K = 10 and W = 30 m. As N increases, the average achievable rate in cluster does not change much. This lack of variation occurs because the minimum rate that determines the rate of the cluster does not change significantly even if N increases. Likewise, the achievable rate improves in the order of LEACH, K-means, and non-SWIPT, and the proposed ENO with PS and TS shows the best performance. On the other hand, as N increases, the total energy dissipated increases, because the total energy dissipated is linearly proportional to N. Nevertheless, the two proposed ENO schemes show significantly lower dissipated energy levels.

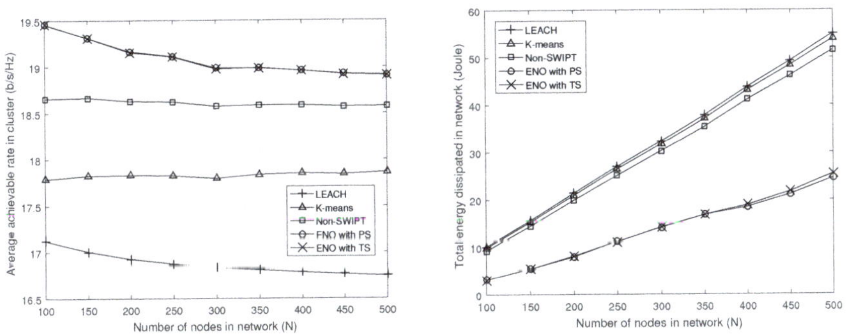

(a) (b)

Figure 6. (**a**) Average achievable rate in cluster and (**b**) total energy dissipated in network vs. number of nodes in network (N) when K = 10 and W = 30 m.

Figure 7 shows the average achievable rate in cluster and the total energy dissipated in network versus the width of the square network (W) when $N = 300$ and $K = 10$. As the network area increases, the average achievable rate decreases because the average distance between nodes increases. On the other hand, the total energy dissipation decreases as the network area increases because the amount of energy harvested in all of the nodes becomes smaller. Likewise, the proposed ENO schemes show better performances than the conventional schemes according to the change in network size.

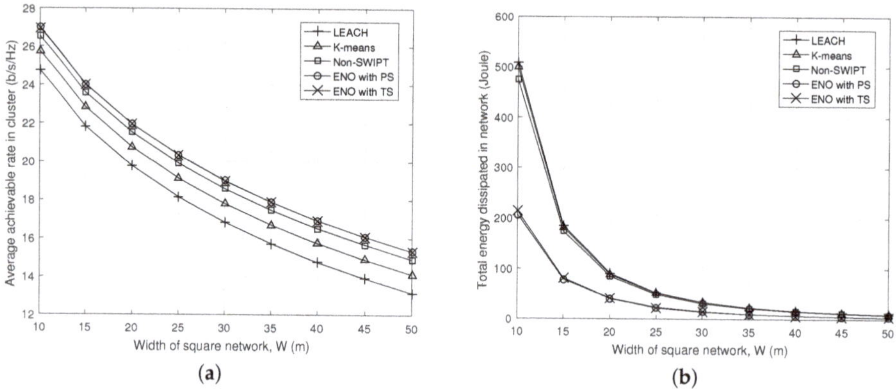

Figure 7. (**a**) Average achievable rate in cluster and (**b**) total energy dissipated in network vs. width in square network (W) when $N = 300$ and $K = 10$.

6. Conclusions

In this study, we proposed a novel ENO framework based on SWIPT in a hierarchical WPSN environment. To maximize the achievable rate of sensing data while guaranteeing ENO, new protocols and algorithms related to the frame structure, ENO, SWIPT ratios, clustering, and CH selection were presented. The simulation results showed that the proposed scheme using SWIPT performs much better in terms of the achievable rate and dissipated energy than the conventional schemes, which do not use SWIPT. It is also evident that the effect of SWIPT is greater when the number of clusters is large and the density of nodes is high. Therefore, we expect that the proposed SWIPT-based ENO can be applied to future WSNs using WPT technologies. In further research, we will investigate the distributed operations of the proposed algorithms and apply other multiple access protocols [48] considering access collision instead of the conflict-free TDMA protocol for more practical operation.

Author Contributions: H.-H.C. contributed to propose the main idea, derive the simulation results, and write most of the paper. J.-R.L. was responsible for mathematical development, verification, and proofreading of the paper.

Funding: This work was supported by the National Research Foundation of Korea (NRF) grants funded by the Korea government (MSIT) (No. 2019R1A2C4070466 and NRF-2019R1F1A1058587).

Acknowledgments: The authors are grateful to the anonymous reviewers for their comments and valuable suggestions.

Conflicts of Interest: The authors declare no conflict of interest.

References

1. Kansal, A.; Hsu, J.; Zahedi, S.; Srivastava, M.B. Power management in energy harvesting sensor networks. *ACM Trans. Embeded Comput. Syst.* **2007**, *6*, 32. [CrossRef]
2. Akan, O.B.; Cetinkaya, O.; Koca, C.; Ozger, M; Internet of hybrid energy harvesting things. *IEEE Internet Things J.* **2017**, *5*, 736–746. [CrossRef]

3. Vigorito, C.M.; Ganesan, D.; Barto, A.G; Adaptive control of duty cycling in energy-harvesting wireless sensor networks. In Proceedings of the 2007 4th Annual IEEE Communications Society Conference on Sensor, Mesh and Ad Hoc Communications and Networks, San Diego, CA, USA, 18–21 June 2007; pp. 21–30.
4. Moser, C.; Thiele, L.; Brunelli, D.; Benini, L. Adaptive power management in energy harvesting systems. In Proceedings of the 2007 Design, Automation & Test in Europe Conference & Exhibition, Nice, France, 16–20 April 2007.
5. Alippi, C.; Anastasi, G.; Di Francesco, M.; Roveri, M. Energy management in wireless sensor networks with energy-hungry sensors. *IEEE Instrum. Meas. Mag.* **2009**, *12*, 16–23. [CrossRef]
6. Eu, Z.A.; Tan, H.P.; Seah, W.K. Opportunistic routing in wireless sensor networks powered by ambient energy harvesting. *Comput. Netw.* **2010**, *54*, 2943–2966. [CrossRef]
7. Nguyen, M.T.; Nguyen, T.H. Wireless Power Transfer: A survey of techniques, and applications on communication networks. *ICSES Trans. Comput. Netw. Commun.* **2018**, *4*, 1–5.
8. Yang, G.; Yuan, D.; Liang, Y.C.; Zhang, R.; Leung, V.C. Optimal resource allocation in full-duplex ambient backscatter communication networks for wireless-powered IoT. *IEEE Internet Things J.* **2018**, *6*, 2612–2625. [CrossRef]
9. Bi, S.; Ho, C.K.; Zhang, R. Wireless powered communication: Opportunities and challenges. *IEEE Commun. Mag.* **2015**, *53*, 117–125. [CrossRef]
10. Sun, Q.; Dai, H.N.; Wang, Q.; Li, X.; Wang, H. When friendly jamming meets wireless energy transfer. In Proceedings of the 2018 IEEE International Conference on Internet of Things (iThings) and IEEE Green Computing and Communications (GreenCom) and IEEE Cyber, Physical and Social Computing (CPSCom) and IEEE Smart Data (SmartData), Halifax, NS, Canada, 30 July–3 August 2018; pp. 320–325.
11. Bi, S.; Zeng, Y.; Zhang, R. Wireless powered communication networks: An overview. *IEEE Wirel. Commun.* **2016**, *23*, 10–18. [CrossRef]
12. Lu, X.; Wang, P.; Niyato, D.; Kim, D.I.; Han, Z. Wireless charging technologies: Fundamentals, standards, and network applications. *IEEE Commun. Surv. Tutor.* **2015**, *18*, 1413–1452. [CrossRef]
13. Liu, Y.; Ding, Z.; Elkashlan, M.; Poor, H.V. Cooperative non-orthogonal multiple access with simultaneous wireless information and power transfer. *IEEE J. Sel. Commun.* **2016**, *34*, 938–953. [CrossRef]
14. Liu, L.; Zhang, R.; Chua, K.C. Wireless information and power transfer: A dynamic power splitting approach. *IEEE Trans. Commun.* **2013**, *61*, 3990–4001. [CrossRef]
15. Sudevalayam, S.; Kulkarni, P. Energy harvesting sensor nodes: Survey and implications. *IEEE Commun. Surv. Tutor.* **2010**, *13*, 443–461. [CrossRef]
16. Yick, J.; Mukherjee, B.; Ghosal D. Wireless sensor network survey. *Comput. Netw.* **2008**, *52*, 2292–2330. [CrossRef]
17. Wang, Q.; Dai, H.N.; Zheng, Z.; Imran, M.; Vasilakos, A. On connectivity of wireless sensor networks with directional antennas. *Sensors* **2017**, *17*, 134. [CrossRef] [PubMed]
18. Singh, S.K.; Singh, M.P.; Singh, D.K. A survey of energy-efficient hierarchical cluster-based routing in wireless sensor networks. *Int. J. Adv. Netw. Appl.* **2010**, *2*, 570–580.
19. Krishnamachari, B.; Estrin, D.; Wicker, S. Modelling data-centric routing in wireless sensor networks. In *IEEE Infocom*; IEEE: New York, NY, USA, 2002; Volume 2, pp. 39–44.
20. Choi, H.H. Construction of energy-efficient data aggregation tree in wireless sensor networks. *J. Korea Inf. Commun. Soc.* **2016**, *41*, 1057–1059. [CrossRef]
21. Ju, H.; Zhang, R. Throughput maximization in wireless powered communication networks. *IEEE Trans. Wireless Commun.* **2013**, *13*, 418–428. [CrossRef]
22. Liu, L.; Zhang, R.; Chua, K.C. Multi-antenna wireless powered communication with energy beamforming. *IEEE Trans. Commun.* **2014**, *62*, 4349–4361. [CrossRef]
23. Chen, H.; Li, Y., Rebelatto, J.L.; Uchoa-Filho, B.F.; Vucetic, B. Harvest-then-cooperate: Wireless-powered cooperative communications. *IEEE Trans. Signal Process.* **2015**, *63*, 1700–1711. [CrossRef]
24. Kang, X.; Ho, C.K.; Sun, S. Full-duplex wireless-powered communication network with energy causality. *IEEE Trans. Wirel. Commun.* **2015**, *14*, 5539–5551. [CrossRef]
25. Yuan, L.; Bi, S.; Zhang, S.; Lin, X.; Wang, H. Multi-antenna enabled cluster-based cooperation in wireless powered communication networks. *IEEE Access* **2017**, *5*, 13941–13950. [CrossRef]
26. Zhang, C.; Zhang, P.; Zhang, W. Cluster cooperation in wireless-powered sensor networks: Modeling and Performance Analysis. *Sensors* **2017**, *17*, 2215. [CrossRef] [PubMed]

27. Yang, Y.; Han, Z.; Ma, G.; Gong, Y.; Qian, L. An energy-efficient hierarchical protocol for wireless powered sensor networks. In Proceedings of the 2018 IEEE 23rd International Conference on Digital Signal Processing (DSP), Shanghai, China, 19–21 November 2018.
28. Lei, M.; Zhang, X.; Ding, H.; Yu, B. Fairness-aware resource allocation in multi-Hop wireless powered communication networks with user cooperation. *Sensors* **2018**, *18*, 1890. [CrossRef] [PubMed]
29. Asiedu, D.K.P.; Shin, S.; Koumadi, K.M.; Lee, K.J. Review of simultaneous wireless information and power transfer in wireless sensor networks. *J. Inf. Commun. Converg. Eng.* **2019**, *17*, 105–116.
30. Hossain, M.A.; Noor, R.M.; Yau, K.L.A.; Ahmedy, I.; Anjum, S.S. Survey on simultaneous wireless information and power transfer with cooperative relay and future challenges. *IEEE Access* **2019**, *7*, 19166–19198. [CrossRef]
31. Perera, T.D.P.; Jayakody, D.N.K.; Sharma, S.K.; Chatzinotas, S.; Li, J. Simultaneous wireless information and power transfer (SWIPT): Recent advances and future challenges. *IEEE Commun. Surv. Tutor.* **2017**, *20*, 264–302. [CrossRef]
32. Guo, S.; Wang, F.; Yang, Y.; Xiao, B. Energy-efficient cooperative tfor simultaneous wireless information and power transfer in clustered wireless sensor networks. *IEEE Trans. Commun.* **2015**, *63*, 4405–4417. [CrossRef]
33. Liu, T.; Wang, X.; Zheng, L. A cooperative SWIPT scheme for wirelessly powered sensor networks. *IEEE Trans. Commun.* **2017**, *65*, 2740–2752. [CrossRef]
34. Tong, B.; Li, Z.; Wang, G.; Zhang, W. How wireless power charging technology affects sensor network deployment and routing. In Proceedings of the 2010 IEEE 30th International Conference on Distributed Computing Systems, Genoa, Italy, 21–25 June 2010; pp. 438–447.
35. Shi, L.; Han, J.; Han, D.; Ding, X.; Wei, Z. The dynamic routing algorithm for renewable wireless sensor networks with wireless power transfer. *Comput. Netw.* **2014**, *74*, 34–52. [CrossRef]
36. Cetinkaya, O.; Dinc, E.; Koca, C.; Merrett, G.V.; Akan, O.B.; Energy-neutral wireless-powered networks. *IEEE Wirel. Commun. Lett.* **2019**. [CrossRef]
37. Park, J.J.; Moon, J.H.; Lee, K.Y.; Kim, D.I. Adaptive mode switching algorithm for dual mode SWIPT with duty cycle operation. In Proceedings of the 2018 IEEE 19th International Workshop on Signal Processing Advances in Wireless Communications (SPAWC), Kalamata, Greece, 25–28 June 2018.
38. Setiawan, D.; Aziz, A.A.; Kim, D.I.; Choi, K.W. Experiment, modeling, and analysis of wireless-powered sensor network for energy neutral power management. *IEEE Syst. J.* **2017**, *12*, 3381–3392. [CrossRef]
39. Choi, K.W.; Ginting, L.; Rosyady, P.A.; Aziz, A.A.; Kim, D.I. Wireless-powered sensor networks: How to realize. *IEEE Trans. Wirel. Commun.* **2016**, *16*, 221–234. [CrossRef]
40. Long, T.; Ozger, M.; Cetinkaya, O.; Akan, O.B. Energy neutral internet of drones. *IEEE Commun. Mag.* **2018**, *56*, 22–28. [CrossRef]
41. Singh, S.K.; Kumar, P.; Singh, J.P. A survey on successors of LEACH protocol. *IEEE Access* **2017**, *5*, 4298–4328. [CrossRef]
42. Yang, Z.; Xu, W.; Pan, Y.; Pan, C.; Chen, M. Optimal fairness-aware time and power allocation in wireless powered communication networks. *IEEE Trans. Commun.* **2018**, *66*, 3122–3135. [CrossRef]
43. Heinzelman, W.R.; Chandrakasan, A.; Balakrishnan, H. Energy-efficient communication protocol for wireless microsensor networks. In Proceedings of the 33rd annual Hawaii international conference on system sciences, Maui, HI, USA, 4–7 January 2000; p. 10.
44. Choi, H.H.; Lee, J.R. Distributed transmit power control for maximizing end-to-end throughput in wireless multi-hop networks. *Wirel. Pers. Commun.* **2014**, *74*, 1033–1044. [CrossRef]
45. Sasikumar, P.; Khara, S. K-means clustering in wireless sensor networks. In Proceedings of the 2012 Fourth International Conference on Computational Intelligence and Communication Networks, Mathura, India, 3–5 November 2012; pp. 140–144.
46. Choi, H.H.; Shin, W. Slotted ALOHA for wireless powered communication networks. *IEEE Access* **2018**, *6*, 53342–53355. [CrossRef]

47. Izadi, D.; Abawajy, J.; Ghanavati, S.; Herawan, T. A data fusion method in wireless sensor networks. *Sensors* **2015**, *15*, 2964–2979. [CrossRef]
48. Yang, Z.; Xu, W.; Pan, Y.; Pan, C.; Chen, M. Energy efficient resource allocation in machine-to-machine communications with multiple access and energy harvesting for IoT. *IEEE Internet Things J.* **2018**, *5*, 229–245. [CrossRef]

energies

MDPI

Article

Energy Efficient Cooperative Computation Algorithm in Energy Harvesting Internet of Things

Haneul Ko [1], Jaewook Lee [2], Seokwon Jang [2], Joonwoo Kim [2] and Sangheon Pack [2,*]

1 Department of Computer Convergence Software, Korea University, Sejong 30019, Korea; heko@korea.ac.kr
2 School of Electrical Engineering, Korea University, Seoul 02841, Korea; iioiioiio123@korea.ac.kr (J.L.); imsoboy2@korea.ac.kr (S.J.); starjoon0202@korea.ac.kr (J.K.)
* Correspondence: shpack@korea.ac.kr

Received: 12 October 2019; Accepted: 22 October 2019; Published: 24 October 2019

Abstract: The limited battery capacity of Internet of Things (IoT) devices is a major deployment barrier for IoT-based computing systems. In this paper, we propose an energy efficient cooperative computation algorithm (EE-CCA). In an EE-CCA, a pair of IoT devices decide whether to offload some parts of the task to the opponent by considering their energy levels and the task deadline. To minimize the energy outage probability while completing most of tasks before their deadlines, we formulate a constraint Markov decision process (CMDP) problem and the optimal offloading strategy is obtained by linear programming (LP). Meanwhile, an optimization problem of finding pairs of IoT devices (i.e., IoT device pairing problem) is formulated under the optimal offloading strategy. Evaluation results demonstrate that the EE-CCA can reduce the energy outage probability up to 78% compared with the random offloading scheme while completing tasks before their deadlines with high probability.

Keywords: offloading; Internet of Things (IoT); energy; constraint Markov decision process (CMDP); optimization

1. Introduction

From the recent advancement of Internet of Things (IoT) devices with high computing power, complicated computation can be handled without remote servers [1]. However, the development speed of batteries for IoT devices is inferior to that of computing module, and thus the limited battery capacity of IoT devices is being a major deployment barrier for IoT-based computing systems. Therefore, there is an increasing interest on the energy harvesting technique that converts wasted energy to electricity [2,3]. With this technique, IoT devices do not need to recharge and/or replace their batteries anymore, and thus the operating expenditure of IoT-based computing systems can be reduced [4]. However, the energy that can be harvested from external energy sources is generally uncontrollable and intermittent. Moreover, the harvested energy volume has temporal and spatial variations. Therefore, it is difficult to provide a reliable power supply to IoT devices. In this situation, if an IoT device cannot harvest energy for a long time and it processes lots of tasks requiring high computing power, its energy can be depleted. To mitigate this problem and improve the energy efficiency of harvesting IoT devices, a number of works (e.g., sleep scheduling, CPU cycle adjustment, and so on) have been investigated in the literature [5–10]. One of the possible solutions is offloading tasks to nearby IoT devices [5–7]. IoT-based computing systems have advantages compared to remote servers-based offloading systems. For example, offloading to remote servers consumes huge resources in networks when IoT devices generate lots of tasks. In addition, longer latency is needed when offloading to remote servers. Especially when an energy-scarce IoT device offloads tasks to a nearby energy-abundant IoT device, energy depletion of the energy-scarce IoT devices probably does not occur. However, unplanned offloading can cause another energy depletion. For example, when an offloader (i.e., an energy-scarce IoT device) always offloads all tasks to an offloadee (i.e., an energy-abundant IoT device) having lots of own tasks and/or small harvesting rate, the energy of the offloadee can be depleted within a short duration. Moreover, tasks cannot be completed within their deadlines due to the high load of the offloadee. Therefore, a sophisticated offloading algorithm should be devised.

In this paper, we propose an energy efficient cooperative computation algorithm (EE-CCA). In an EE-CCA, each IoT device is paired to its partner and a pair of IoT devices conduct cooperative computing. Specifically, a centralized controller collects information such as the distribution about temporal and spatial variations of external energy sources, the task occurrence rates of IoT devices, and the energy levels of IoT devices. Based on this information, the controller can construct and distribute offloading decision tables to IoT devices. Then, when a task occurs in IoT devices, they decide whether to offload some parts of the task to the opponent by following the decision tables. To minimize the energy outage probability while completing most of the tasks before their deadlines, we formulate a constraint Markov decision process (CMDP) problem, and the optimal offloading strategy is obtained by linear programming (LP). Meanwhile, an optimization problem of finding pairs of IoT devices (i.e., IoT device pairing problem) is formulated under the optimal offloading strategy. Evaluation results demonstrate that the EE-CCA can reduce the energy outage probability up to 78% compared with the random offloading scheme while completing tasks before their deadlines with high probability. In addition, the EE-CCA operates adaptively even when the operating environment (e.g., inter-task occurrence rate) changes.

The contribution of this paper can be summarized as follows: (1) we develop the cooperative computation algorithm called EE-CCA for IoT devices, while optimizing the EE-CCA by means of CMDP formulation; (2) optimal pairs of IoT devices are decided based on the optimization problem; and (3) extensive evaluation results are presented and analyzed under various environments, providing valuable guidelines for the design of cooperative computing in energy harvesting IoT.

The remainder of this paper is organized as follows. Related works are summarized in Section 2, and the EE-CCA is described in Section 3. The CMDP model for cooperative computing and the optimization problem for the IoT device pairing are developed in Section 4. Evaluation results are given in Section 5, and followed by the concluding remarks in Section 6.

2. Related Works

A number of studies on the computation offloading have been conducted to improve the energy efficiency of energy-constraint devices [11–22]. These can be categorized into: (1) framework design [11–16]; and (2) algorithm design [17–22].

Wang et al. [11] proposed an evolutionary mobile network architecture called MobiScud that integrates the cloud services into the mobile networks by means of software defined networking (SDN) and network function virtualization (NFV) technologies in a backwards compatible fashion. Tong et al. [12] organized edge cloud servers into a hierarchical architecture which enables aggregation of the peak loads across different tiers of cloud servers. Specifically, when the loads exceed the capacities of lower tiers of edge cloud servers, they can be aggregated and offloaded by other servers at higher tiers in the edge cloud hierarchy to maximize the amount of mobile workloads being served. Taleb and Ksentini [13] introduced a follow me cloud concept that enables mobile cloud services to follow their respective mobile devices by migrating services to the optimal cloud. Liu et al. [14] proposed convergence of cloud and cellular systems, abbreviated as CONCERT, based on a concept of control/data plane decoupling and hierarchically placement of the resources within the network to manage flexibly and elastically networks and cloud services. Puente et al. [15] presented a seamless approach for the deployment of edge clouds where conventional mobile traffic and computing related traffic are segregated and handled individually at base stations. However, since these works do not consider the device-to-device offloading, IoT devices with high computing power cannot be exploited efficiently. Shukla and Munir [16] proposed a computation offloading architecture where an IoT device first tries to offload tasks to another IoT device instead of directly offloading to the cloud to process the huge amount of data while guaranteeing the task completion before the deadline. However, they did not provide any optimization method.

Ko et al. [17] proposed a spatial and temporal computation offloading decision algorithm where an energy-constraint device decides where and when to process tasks by means of a Markov decision process (MDP) by considering the energy consumption and the transmission cost. Zhao et al. [18] developed an optimization problem whose objective function is to maximize the probability that task execution satisfies the given delay bound. The problem was proved to be concave, and an optimal algorithm was proposed. Tang and Chen [19] studied a social-aware computation offloading game and designed a distributed computation offloading algorithm to achieve the Nash equilibrium. Similarly, Chen et al. [20] modeled a multi-user computation offloading game and designed a distributed computation offloading algorithm that can achieve the Nash equilibrium of the game. Zheng et al. [21] formulated the mobile users' offloading decision process under a dynamic environment as a stochastic game. Then, they proposed a multi-agent stochastic learning algorithm that can run in a fully distributed manner without any information exchange. Yu et al. [22] developed an optimal collaborative offloading strategy under a distributed caching scenario. Specifically, they formulated a problem of users' allocation as a coalition formation game with the consideration of relationships between the offloading and caching and then proposed an optimal offloading with a caching-enhancement scheme. These works improve the performance of computation offloading; however, no previous studies optimize the performance of IoT-based computing systems.

3. Energy Efficient Cooperative Computation Algorithm (EE-CCA)

Figure 1 shows the system model of this paper. In our system model, there are N IoT devices with the energy harvesting capability. We assume heterogeneous IoT environment, where IoT devices have different computing power and current energy level. In real systems, some IoT devices may not have sufficient computing power and/or harvesting capability. To support this situation, a study on the robustness of the proposed algorithm (e.g., a resilient multiscale coordination control [23]) should be conducted, which is one of our future works. In addition, since they are installed at different spots and the condition of external energy sources is volatile, energy volumes that can be harvested at each IoT device are different from each other. In addition, with different rates, these IoT devices periodically generate tasks that can be abstracted into the input data and the completion deadline before which the task should be completed [24]. In addition, we consider an application where the input data do not have dependency, e.g., binomial classification that determines whether each input is larger than a given threshold or not, and therefore the input data can be partitioned and offloaded. When the task occurs in a particular IoT device, it decides whether to offload some parts of the task (i.e., input data) to a neighbor IoT device or not with the consideration of the energy level and the deadline of the task. Note that, even though the formulation of this paper is based on the assumption where IoT devices can offload all or half of the tasks, it can be easily extended to consider other portions of the task.

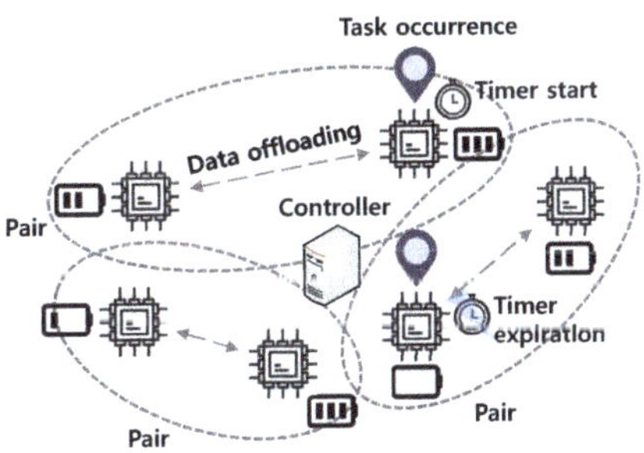

Figure 1. System model.

Intuitively, when an IoT device offloads some parts of the task to a neighbor IoT device, the task can be processed in a distributed manner, which can reduce the energy consumption of the task owner. Since IoT devices offload their tasks to nearby IoT devices by exploiting transmission technologies with low power consumption (e.g., Bluetooth), the energy consumption for transmission of the task can be neglected compared to that for processing the task. Moreover, if the neighbor IoT device does not have its own task, the task completion time can be shortened. However, if the task is offloaded to an energy-scarce IoT device, it causes the energy depletion of the IoT device, and then the offloaded task cannot be processed due to the energy depletion. In addition, when the neighbor IoT device has its own task, it should process its own task and the offloaded task simultaneously, and thus its processing time can increase. Then, both tasks may not be completed within their deadlines. To prevent these situations, we propose the EE-CCA and its flow chart is shown in Figure 2. First, the controller collects and/or maintains information such as the distribution about temporal and spatial variations of external energy sources, the task occurrence rates of IoT devices, and the energy levels of IoT devices (Step 1 in Figure 2). Based on this information, the controller determines appropriate partners of IoT devices for cooperative computation and transmits the pairing information to IoT devices (Step 2 in Figure 2). In addition, the controller constructs an offloading decision table consisting of the current status and the operation in a centralized manner. If some parameters (e.g., the task occurrence rates of IoT devices) are changed, the offloading decision table can be reconstructed by the controller and transmitted to IoT devices again. Therefore, if the parameters are frequently changed, extra signaling overhead can occur. To mitigate the signaling overhead, several techniques such as aggregation and delta encoding can be exploited [25]. Note that the offloading decision table can be obtained by CMDP, which will be elaborated in Section 4. After that, the controller transmits the optimal offloading decision table to IoT devices (Step 3 in Figure 2). On the basis of this table, IoT devices can conduct the cooperative computing (i.e., decide whether to offload some parts of the task or not) (Step 4 in Figure 2). By means of table deployments in IoT devices, the CMDP model can be applied to resource-constrained IoT devices without any high computation overhead in IoT devices [26].

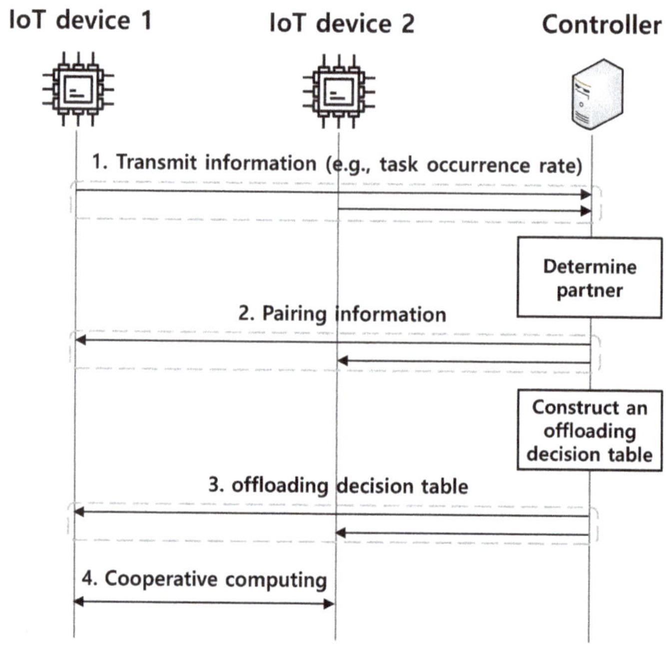

Figure 2. Flow chart.

4. Constraint Markov Decision Process (CMDP)

To obtain the optimal offloading strategy, we formulate a CMDP model with five elements. (Since the CMDP model that is a mathematical framework to model decision-making when outcomes need to be constrained, and they are partially random and under the control of the decision maker [27], it is suitable to construct the optimal offloading decision table.): (1) decision epoch; (2) state; (3) action; (4) transition probability; and (5) cost and constraint functions. Subsequently, we convert the CMDP model to an equivalent LP problem to obtain the optimal policy. After that, the IoT device pairing problem is formulated under the optimal offloading policy. Important notations for the CMDP model and IoT device pairing problem are summarized in Table 1.

Table 1. Summary of notations.

Notation	Description
S_t	State at the decision epoch t
A_t	Action chosen at the decision epoch t
τ	Duration of each decision epoch
$\mathbf{S}$	Overall state space
$\mathbf{T_i^M}$	State for denoting the occurrence and processing status for the task of IoT device i
$\mathbf{T_j^M}$	State for denoting the occurrence and processing status for the task of IoT device j
$\mathbf{T_i^O}$	State for denoting the processing status for the offloaded task of IoT device i
$\mathbf{T_j^O}$	State for denoting the processing status for the offloaded task of IoT device j
$\mathbf{E_i}$	State for denoting the energy level of IoT device i
$\mathbf{E_j}$	State for denoting the energy level of IoT device j
$\mathbf{D_i}$	State for denoting whether the timer for the deadline of the task of IoT device i expires or not
$\mathbf{D_j}$	State for denoting whether the timer for the deadline of the task of IoT device j expires or not
$\mathbf{A}$	Action space
$\mathbf{A_i}$	Action space for IoT device i
$\mathbf{A_j}$	Action space for IoT device j
E_{MAX}	Maximum battery capacity of IoT device
$r(S,A)$	Cost function on the energy outage
$c_i(S,A)$	Constraint function on the timer expiration of IoT device i
$c_j(S,A)$	Constraint function on the timer expiration of IoT device j
ζ^E	Energy outage probability
ζ_i^T	Timer expiration probability of IoT device i
ζ_j^T	Timer expiration probability of IoT device j
θ_i^T	Upper limit on the timer expiration probability of IoT device i
θ_j^T	Upper limit on the timer expiration probability of IoT device j
ζ_{ij}^E	Individual energy outage probability of IoT device i when it is paired with IoT device j
x_{ij}	Decision variable to denote whether IoT device i is paired with IoT device j or not

4.1. Decision Epoch

Figure 3 shows the timing diagram for the CMDP model. A sequence $T = \{1,2,3,...\}$ represents the time epochs when successive decisions are made [28]. S_t and A_t denote the state and the action chosen at the decision epoch $t \in T$, respectively. τ represents the duration of each decision epoch.

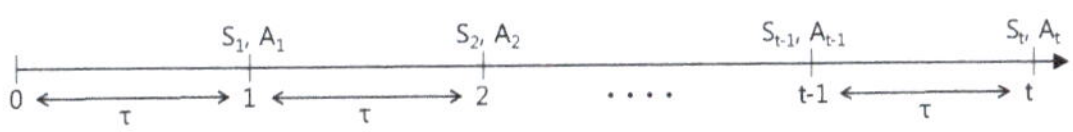

Figure 3. Timing diagram.

4.2. State Space

We define the overall state space $\mathbf{S}$ as (The state space is constructed based on the assumption where IoT devices can offload all or half of the task. However, it can be easily extended to consider other portions of the task to add elements to $\mathbf{T_i^M}$, $\mathbf{T_i^O}$, $\mathbf{T_j^M}$, and $\mathbf{T_j^O}$.)

$$\mathbf{S} = \mathbf{T_i^M} \times \mathbf{T_i^O} \times \mathbf{E_i} \times \mathbf{D_i} \times \mathbf{T_j^M} \times \mathbf{T_j^O} \times \mathbf{E_j} \times \mathbf{D_j}, \tag{1}$$

where $\mathbf{T_i^M}$ and $\mathbf{T_j^M}$ are the states for representing the occurrence and processing status for the task of IoT devices i and j, respectively. $\mathbf{T_i^O}$ and $\mathbf{T_j^O}$ denote the states for the processing status of the offloaded task of IoT devices i and j, respectively. $\mathbf{E_i}$ and $\mathbf{E_j}$ are the states for the energy level of IoT devices i and j, respectively. $\mathbf{D_i}$ and $\mathbf{D_j}$ represent the states for denoting whether the timers for the deadline of the task of IoT devices i and j expire or not, respectively.

$\mathbf{T_i^M}$, $\mathbf{T_i^O}$, $\mathbf{E_i}$, and $\mathbf{D_i}$ are the states for IoT device i, and these states are defined as follows.

First, $\mathbf{T_i^M}$ is given by

$$\mathbf{T_i^M} = \{0, 1, 2, 3, 4\}, \tag{2}$$

where T_i^M represents the occurrence and processing status for the task of IoT device i. In other words, $T_i^M = 0$ denotes that the task does not occur in IoT device i, whereas $T_i^M = 1$ refers to the situation immediately after the task occurs in IoT device i. $T_i^M = 2$ represents the situation where IoT device i processes all of the task by itself. Meanwhile, $T_i^M = 3$ and $T_i^M = 4$ represent the situations where half of the task and all of the tasks are offloaded to IoT device j, respectively. Note that, when $T_i^M = 3$, IoT device i processes the remaining half of the task.

$\mathbf{T_i^O}$ is represented by

$$\mathbf{T_i^O} = \{0, 1, 2\}, \tag{3}$$

where T_i^O represents the processing status of the offloaded task of IoT device i. Specifically, $T_i^O = 0$ describes the situation where any task is not offloaded to its partner (i.e., IoT device j). Meanwhile, $T_i^O = 1$ and $T_i^O = 2$ represent the situation where half of the task and all of the tasks are offloaded to IoT device j, respectively, and it is being processed in IoT device j.

$\mathbf{E_i}$ is represented as [29]

$$\mathbf{E_i} = \{0, 1, 2, ... E_{MAX}\}, \tag{4}$$

where E_{MAX} is the maximum battery capacity of an IoT device.

$$\mathbf{D_i} = \{0, 1\}, \tag{5}$$

where D_i denotes whether the timer for the deadline of the task of IoT device i expires or not. In other words, $D_i = 0$ and $D_i = 1$ represent that the timer for the deadline of the task of IoT device i does not expire and expires, respectively.

$\mathbf{T_j^M}$, $\mathbf{T_j^O}$, $\mathbf{E_j}$, and $\mathbf{D_j}$ are the states for IoT device j, and these states can be defined as similar with the states for IoT device i. These definitions are omitted in this paper due to the page limitation and for simple descriptions, which can be found in [30].

4.3. Action Space

When the task occurs in IoT devices, each IoT device can decide whether to offload to its partner or not and the portion to be offloaded based on the current state information. The action set is constructed based on the assumption where IoT devices can offload all or half of the task. However, it can be easily extended to consider other portions of the task to define additional actions. Therefore, the action set can be described by

$$\mathbf{A} = \mathbf{A_i} \times \mathbf{A_j}, \tag{6}$$

where $\mathbf{A_i}$ and $\mathbf{A_j}$ are the action spaces for IoT devices i and j, respectively, which can be defined as

$$\mathbf{A_i} = \{0, 1, 2\} \tag{7}$$

and

$$\mathbf{A_j} = \{0, 1, 2\}, \tag{8}$$

where $A_i = 0$ and $A_j = 0$ represent that IoT devices i and j do not offload its task, respectively. $A_i = 1$ and $A_j = 1$ denote that IoT devices i and j offload half of the task to its partner, respectively. In addition, $A_i = 2$ and $A_j = 2$ are the actions where IoT devices i and j offload all of the task to its partner, respectively.

4.4. Transition Probability

The state transition probability of IoT device i is affected by the state of IoT device j. Specifically, the processing speed of the task occurring in IoT device i (i.e., the transition probability of $\mathbf{T_i^M}$) is dependent on whether the task occurring at IoT device j is processed in IoT device i or not (i.e., T_j^O). In addition, the transition probability of T_i^O is affected by whether IoT device j processes its own task or not (i.e., T_j^M). Similar to that, the state transition probability of IoT device j is also influenced by the state of IoT device i (especially T_i^M and T_i^O). Therefore, the transition probability with the chosen action A from the current state S to the next state S' can be described by

$$P[S'|S, A] = P[S_i'|S_i, T_j^M, T_j^O, A] \times P[S_j'|S_j, T_i^M, T_i^O, A], \tag{9}$$

where S_i' and S_j' denote the next state of IoT devices i and j, respectively. In addition, S_i and S_j represent the current state for IoT devices i and j, respectively.

Meanwhile, T_i^M and T_i^O are influenced by the chosen action A, and these states are dependently changed with each other. In addition, T_i^M is affected by T_j^O. For example, when the task of IoT device j is processed in IoT device i, the processing speed of the task of IoT device i can decrease. Similarly, T_i^O is influenced by T_j^M. For example, when IoT device j does not process its own task, it can focus on processing the offloaded task from IoT device i, and therefore the offloaded task can be completed within a short duration. Meanwhile, when the task is processed in IoT device i, its energy level can decrease. That is, the transition of E_i is influenced by T_i^M. The timer for the deadline of the task operates only when the task occurs, and therefore the transition of D_i is affected by T_i^M and T_i^O. Meanwhile, other states change independently of each other. Therefore, for the chosen action A, the transition probability from the current state of IoT device i, $S_i = [T_i^M, T_i^O, E_i, D_i]$, to the next state of IoT device i, $S_i' = [T'^M_i, T'^O_i, E_i, D_i]$, can be described by

$$P[S_i'|S_i, T_j^M, T_j^O, A] = P[T'^M_i|T_i^M, T_i^O, T_j^O, A] \times P[T'^O_i|T_i^O, T_i^M, T_j^M, A] \times P[E_i'|E_i, T_i^M] \times P[D_i'|D_i, T_i^M, T_i^O]. \tag{10}$$

We assume that the inter-task occurrence time of IoT device i follows an exponential distribution with mean $1/\lambda_i$. Then, the probability that the task occurs in IoT device i during a decision epoch can be calculated as $\lambda_i\tau$ [27,31]. Therefore, $P[T'^M_i|T^M_i = 0, T^O_i = 0, A]$ can be represented by

$$P[T'^M_i|T^M_i = 0, T^O_i = 0, T^O_j, A] = \begin{cases} 1 - \lambda_i\tau, & \text{if } T'^M_i = 0, \\ \lambda_i\tau, & \text{if } T'^M_i = 1, \\ 0, & \text{otherwise.} \end{cases} \tag{11}$$

Before receiving the result of the offloaded task, IoT device i does not generate the task. Therefore, $P[T'^M_i|T^M_i = 0, T^O_i \neq 0, T^O_j, A]$ can be defined as

$$P[T'^M_i|T^M_i = 0, T^O_i \neq 0, T^O_j, A] = \begin{cases} 1, \text{ if } T'^M_i = 0, \\ 0, \text{ otherwise.} \end{cases} \tag{12}$$

Meanwhile, when the task occurs (i.e., $T^M_i = 1$), IoT device i decides whether to offload to IoT device j or not and the offloaded portion (i.e., half of the task and all of the task). If IoT device i decides not to offload it (i.e., when $A = 0$), the task state will change to 2 representing the situation where IoT device i processes all of the task by itself (i.e., $T^M_i = 2$). On the other hand, when IoT device i decides to offload half of the task and all of the task (i.e., when $A = 1$ and $A = 2$), the next states of the occurrence and processing status for the task of IoT device i (i.e., T'^M_i) become 3 and 4, respectively. Therefore, the corresponding transition probabilities can be represented as

$$P[T'^M_i|T^M_i = 1, T^O_i, T^O_j, A = 0] = \begin{cases} 1, \text{ if } T'^M_i = 2, \\ 0, \text{ otherwise,} \end{cases} \tag{13}$$

$$P[T'^M_i|T^M_i = 1, T^O_i, T^O_j, A = 1] = \begin{cases} 1, \text{ if } T'^M_i = 3, \\ 0, \text{ otherwise,} \end{cases} \tag{14}$$

and

$$P[T'^M_i|T^M_i = 1, T^O_i, T^O_j, A = 2] = \begin{cases} 1, \text{ if } T'^M_i = 4, \\ 0, \text{ otherwise,} \end{cases} \tag{15}$$

respectively.

We assume that the processing time of IoT device i for its own task follows an exponential distribution with mean $1/\mu^{F,S}_i$ when it processes all of its own task and any task of IoT device j is not offloaded to IoT device i (i.e., $T^M_i = 2$ and $T^O_j = 0$). In this case, the probability that the task is completed during a decision epoch is $\mu^{F,S}_i\tau$ [27,31]. Then, the probability that a task is not completed during a decision epoch is $1 - \mu^{F,S}_i\tau$. On the other hand, if some portion of the task of IoT device j is offloaded to IoT device i (i.e., $T^O_j \neq 0$), the processing speed of IoT device i for its own task decreases, and thus it is assumed that the processing time of IoT device i for its own task follows an exponential distribution with mean $1/\mu^{F,D}_i$ ($> 1/\mu^{F,S}_i$). In this case, the probability that the task is completed (or not completed) during a decision epoch is $\mu^{F,D}_i\tau$ (or $1 - \mu^{F,D}_i\tau$) [27,31]. Therefore, $P[T'^M_i|T^M_i = 2, T^O_i, T^O_j = 0, A]$ and $P[T'^M_i|T^M_i = 2, T^O_i, T^O_j \neq 0, A]$ can be derived as

$$P[T'^M_i|T^M_i = 2, T^O_i, T^O_j = 0, A] = \begin{cases} 1 - \mu^{F,S}_i\tau, & \text{if } T'^M_i = 2, \\ \mu^{F,S}_i\tau, & \text{if } T'^M_i = 0, \\ 0, & \text{otherwise,} \end{cases} \tag{16}$$

and

$$P[T'^M_i | T^M_i = 2, T^O_i, T^O_j \neq 0, A] = \begin{cases} 1 - \mu_i^{F,D}\tau, & \text{if } T'^M_i = 2, \\ \mu_i^{F,D}\tau, & \text{if } T'^M_i = 0, \\ 0, & \text{otherwise.} \end{cases} \quad (17)$$

Meanwhile, when offloading half of the task to IoT device j (i.e., $T^M_i = 3$), the remained task can be completed with shorter time. It is assumed that the processing time of the remained task follows an exponential distribution with mean $1/\mu_i^{H,S}$ when any task of IoT device j is not offloaded to IoT device i (i.e., $T^O_j = 0$), and then the probability that the remained task is completed during a decision epoch is $\mu_i^{H,S}\tau$ [27,31]. On the other hand, when IoT device i offloads the half of its task to IoT device j and processes the offloaded task from IoT device j (i.e., $T^M_i = 3$ and $T^O_j \neq 0$), the processing time of the remained task of IoT device i follows an exponential distribution with mean $1/\mu_i^{H,D}$. In this case, the probability that the remained task is completed during a decision epoch is $\mu_i^{H,D}\tau$ [27,31]. Thus, the corresponding transition probabilities can be represented as

$$P[T'^M_i | T^M_i = 3, T^O_i, T^O_j = 0, A] = \begin{cases} 1 - \mu_i^{H,S}\tau, & \text{if } T'^M_i = 3, \\ \mu_i^{H,S}\tau, & \text{if } T'^M_i = 0, \\ 0, & \text{otherwise,} \end{cases} \quad (18)$$

and

$$P[T'^M_i | T^M_i = 3, T^O_i, T^O_j \neq 0, A] = \begin{cases} 1 - \mu_i^{H,D}\tau, & \text{if } T'^M_i = 3, \\ \mu_i^{H,D}\tau, & \text{if } T'^M_i = 0, \\ 0, & \text{otherwise.} \end{cases} \quad (19)$$

When the task does not occur, the processing status of the offloaded task of IoT device i does not change. Therefore, $P[T'^O_i | T^O_i = 0, T^M_i \neq 1, A]$ can be denoted as

$$P[T'^O_i | T^O_i = 0, T^M_i \neq 1, T^M_j, A] = \begin{cases} 1, & \text{if } T'^O_i = 0, \\ 0, & \text{otherwise.} \end{cases} \quad (20)$$

Meanwhile, when the task occurs (i.e., $T^M_i = 1$), the processing status of the offloaded task of IoT device i changes according to the chosen action A. Therefore, corresponding transition probabilities can be represented as

$$P[T'^O_i | T^O_i = 0, T^M_i = 1, T^M_j, A = 0] = \begin{cases} 1, & \text{if } T'^O_i = 0, \\ 0, & \text{otherwise,} \end{cases} \quad (21)$$

$$P[T'^O_i | T^O_i = 0, T^M_i = 1, T^M_j, A = 1] = \begin{cases} 1, & \text{if } T'^O_i = 1, \\ 0, & \text{otherwise,} \end{cases} \quad (22)$$

and

$$P[T'^O_i | T^O_i = 0, T^M_i = 1, T^M_j, A = 2] = \begin{cases} 1, & \text{if } T'^O_i = 2, \\ 0, & \text{otherwise,} \end{cases} \quad (23)$$

respectively.

When some portion of the task is offloaded (i.e., $T^O_i = 1$ or $T^O_i = 2$), it is processed by IoT device j. Meanwhile, the processing time of the offloaded task to IoT device j depends on the portion of the offloaded task (i.e., T_i^O) and whether IoT device j processes its own task or not (i.e., T_j^M). Specifically, when half of the task (or all of the task) of IoT device i is offloaded and IoT device j does not process its own task, we assume that the processing time of the offloaded task follows an exponential distribution with mean $1/\mu_j^{H,S}$ ($1/\mu_j^{F,S}$). On the other hand, if half of the task (or all of the task) of IoT device i is offloaded and IoT device j processes its own task, the processing time of the offloaded task follows an exponential distribution with mean $1/\mu_j^{H,D}$ ($1/\mu_j^{F,D}$). Then, the probabilities that the offloaded task is completed during a decision epoch for each case can be derived as $\mu_j^{H,S}\tau$, $\mu_j^{F,S}\tau$, $\mu_j^{H,D}\tau$, and $\mu_j^{F,D}\tau$, respectively [27,31]. Therefore, the corresponding transition probabilities can be denoted as

$$P[T'^O_i|T^O_i = 1, T^M_i, T^M_j = 0, A] = \begin{cases} 1-\mu_j^{H,S}\tau, & \text{if } T'^O_i = 1, \\ \mu_j^{H,S}\tau, & \text{if } T'^O_i = 0, \\ 0, & \text{otherwise}, \end{cases} \tag{24}$$

$$P[T'^O_i|T^O_i = 2, T^M_i, T^M_j = 0, A] = \begin{cases} 1-\mu_j^{F,S}\tau, & \text{if } T'^O_i = 1, \\ \mu_j^{F,S}\tau, & \text{if } T'^O_i = 0, \\ 0, & \text{otherwise}, \end{cases} \tag{25}$$

$$P[T'^O_i|T^O_i = 1, T^M_i, T^M_j \neq 0, A] = \begin{cases} 1-\mu_j^{H,D}\tau, & \text{if } T'^O_i = 1, \\ \mu_j^{H,S}\tau, & \text{if } T'^O_i = 0, \\ 0, & \text{otherwise}, \end{cases} \tag{26}$$

and

$$P[T'^O_i|T^O_i = 2, T^M_i, T^M_j \neq 0, A] = \begin{cases} 1-\mu_j^{F,D}\tau, & \text{if } T'^O_i = 1, \\ \mu_j^{H,S}\tau, & \text{if } T'^O_i = 0, \\ 0, & \text{otherwise}. \end{cases} \tag{27}$$

The IoT device can harvest energy only when its environments provide energy (e.g., when the wind blows above a certain speed). Therefore, the probability that IoT device i harvests one unit energy at an arbitrary decision epoch is modeled by a Bernoulli random process with the probability p_i^H [32]. Then, when the IoT device i does not process any task (i.e., $T_i^M = 0$ or $T_i^M = 1$) and its battery is not fully charged (i.e., $E_i \neq E_{MAX}$), E_i increases by one unit with the probability p_i^H. If the battery of IoT device i is full (i.e., $E_i = E_{MAX}$), it cannot harvest energy anymore. Therefore, the corresponding transition probabilities can be represented as

$$P[E'_i|E_i \neq E_{MAX}, T^M_i = 0] = \begin{cases} 1-p_i^H, & \text{if } E'_i = E_i, \\ p_i^H, & \text{if } E'_i = E_i + 1, \\ 0, & \text{otherwise}, \end{cases} \tag{28}$$

$$P[E'_i|E_i = E_{MAX}, T^M_i = 0] = \begin{cases} 1, & \text{if } E'_i = E_i, \\ 0, & \text{otherwise}, \end{cases} \tag{29}$$

$$P[E'_i|E_i \neq E_{MAX}, T^M_i = 1] = \begin{cases} 1-p_i^H, & \text{if } E'_i = E_i, \\ p_i^H, & \text{if } E'_i = E_i + 1, \\ 0, & \text{otherwise}, \end{cases} \tag{30}$$

and

$$P[E_i'|E_i = E_{MAX}, T_i^M = 1] = \begin{cases} 1, \text{ if } E_i' = E_i, \\ 0, \text{ otherwise.} \end{cases} \tag{31}$$

When IoT device i processes the task (i.e., $T_i^M = 2$ or $T_i^M = 3$) and it has energy (i.e., $E \neq 0$), it consumes one unit energy. On the other hand, if IoT device i does not have any energy, it cannot process for any task the sensed data, and thus no energy is consumed. In addition, its energy E_i increases by one unit with the probability p_i^H. Therefore, the corresponding transition probabilities can be expressed by

$$P[E_i'|E_i \neq 0, T_i^M = 2] = \begin{cases} 1 - p_i^H, & \text{if } E_i' = E_i - 1, \\ p_i^H, & \text{if } E_i' = E_i, \\ 0, & \text{otherwise,} \end{cases} \tag{32}$$

$$P[E_i'|E_i = 0, T_i^M = 2] = \begin{cases} 1, \text{ if } E_i' = E_i, \\ 0, \text{ otherwise,} \end{cases} \tag{33}$$

$$P[E_i'|E_i \neq 0, T_i^M = 3] = \begin{cases} 1 - p_i^H, & \text{if } E_i' = E_i - 1, \\ p_i^H, & \text{if } E_i' = E_i, \\ 0, & \text{otherwise,} \end{cases} \tag{34}$$

and

$$P[E_i'|E_i = 0, T_i^M = 3] = \begin{cases} 1, \text{ if } E_i' = E_i, \\ 0, \text{ otherwise.} \end{cases} \tag{35}$$

Meanwhile, when all of the tasks are offloaded to IoT device j (i.e., $T_i^M = 4$), IoT device i does not consume its own energy. Therefore, the corresponding transition probability can be denoted as

$$P[E_i'|E_i, T_i^M = 4] = \begin{cases} 1, \text{ if } E_i' = E_i, \\ 0, \text{ otherwise.} \end{cases} \tag{36}$$

When the task does not occur (i.e., $T_i^M = 0$ or $T_i^O = 0$), the timer for the deadline of the task does not start, and therefore it does not expire. Therefore, $P[D_i'|D_i, T_i^M = 0, T_i^O = 0]$ can be represented as

$$P[D_i'|D_i, T_i^M = 0, T_i^O = 0] = \begin{cases} 1, \text{ if } D_i' = 0, \\ 0, \text{ otherwise.} \end{cases} \tag{37}$$

We assume that the timer for the deadline of the task of IoT device i follows an exponential distribution with mean $1/\kappa_i$ [33,34]. Then, when the task is not completed (i.e., $T_i^M \neq 0$ or $T_i^O \neq 0$), the probability that the timer expires during a decision epoch is $\kappa_i \tau$. Thus, the corresponding transition probabilities can be represented as

$$P[D_i'|D_i = 0, T_i^M \neq 0, T_i^O] = \begin{cases} 1 - \kappa_i \tau, & \text{if } D_i' = 0, \\ \kappa_i \tau, & \text{if } D_i' = 1, \\ 0, & \text{otherwise,} \end{cases} \tag{38}$$

and

$$P[D_i'|D_i = 0, T_i^M, T_i^O \neq 0] = \begin{cases} 1-\kappa_i\tau, & \text{if } D_i' = 0, \\ \kappa_i\tau, & \text{if } D_i' = 1, \\ 0, & \text{otherwise.} \end{cases} \quad (39)$$

Meanwhile, when the task is completed (i.e., $T_i^M = 0$ and $T_i^O = 0$), the timer is reset and does not operate, which means that there is no expiration. Therefore, $P[D_i{}'|D_i = 1, T_i^M = 0, T_i^O = 0]$ can be denoted as

$$P[D_i'|D_i = 1, T_i^M = 0, T_i^O = 0] = \begin{cases} 1, & \text{if } D_i' = 0, \\ 0, & \text{otherwise.} \end{cases} \quad (40)$$

If the timer expires (i.e., $D_i = 1$) and the task is not completed (i.e., $T_i^M \neq 0$ or $T_i^O \neq 0$), the timer remains in the expired state. Therefore, the corresponding transition probabilities can be represented as

$$P[D_i'|D_i = 1, T_i^M \neq 0, T_i^O] = \begin{cases} 1, & \text{if } D_i' = 1, \\ 0, & \text{otherwise,} \end{cases} \quad (41)$$

and

$$P[D_i{}'|D_i = 1, T_i^M, T_i^O \neq 0] = \begin{cases} 1, & \text{if } D_i' = 1, \\ 0, & \text{otherwise.} \end{cases} \quad (42)$$

The transition probability for the states of IoT device j can be defined as similar to that of IoT device i. These are omitted in this paper due to the page limitation and for simple descriptions, which can be found in [30].

4.5. Cost and Constraint Functions

4.5.1. Cost Function

To define the cost function, we consider the energy outage of IoT devices. The energy outage occurs when batteries of IoT devices are empty. Therefore, the cost function can be defined as

$$r(S, A) = \begin{cases} 1, & \text{if } E_i = 0, \\ 1, & \text{if } E_j = 0, \\ 0, & \text{otherwise.} \end{cases} \quad (43)$$

4.5.2. Constraint Function

To prevent the situation where the task cannot be finished before the timer expiration, the constraint functions for the timer expiration of IoT devices i and j can be represented by

$$c_i(S, A) = \begin{cases} 1, & \text{if } D_i = 1, \\ 0, & \text{otherwise,} \end{cases} \quad (44)$$

and

$$c_j(S, A) = \begin{cases} 1, & \text{if } D_j = 1, \\ 0, & \text{otherwise,} \end{cases} \quad (45)$$

respectively.

4.6. Optimization Problem Formulation

Since $\sum_{t'}^{t} \mathrm{E}\left[r\left(S_{t'}, A_{t'}\right)\right]$ means the number of energy outages, the average energy outage probability ζ^E can be defined as

$$\zeta^E = \lim_{t\to\infty} \sup \frac{1}{t} \sum_{t'}^{t} \mathrm{E}\left[r\left(S_{t'}, A_{t'}\right)\right], \tag{46}$$

where lim denotes the value that a function approaches as the input approaches a specific value. In addition, sup (i.e., supremum) means the least upper bound.

Meanwhile, the average timer expiration probabilities of IoT devices i and j, denoted as, ζ_i^T, and ζ_j^T, respectively, can be defined as

$$\zeta_i^T = \lim_{t\to\infty} \sup \frac{1}{t} \sum_{t'}^{t} \mathrm{E}\left[c_i\left(S_{t'}, A_{t'}\right)\right] \tag{47}$$

and

$$\zeta_j^T = \lim_{t\to\infty} \sup \frac{1}{t} \sum_{t'}^{t} \mathrm{E}\left[c_j\left(S_{t'}, A_{t'}\right)\right]. \tag{48}$$

Then, the optimization problem in the CMDP model can be formulated as

$$\min_{\pi} \zeta^E, \tag{49}$$

$$\text{s.t.} \quad \zeta_i^T \leq \theta_i^T \ \text{ and } \ \zeta_j^T \leq \theta_j^T, \tag{50}$$

where θ_i^T and θ_j^T are the upper limits on the timer expiration probabilities of IoT devices i and j, respectively.

The formulated optimization problem can be transformed into an equivalent LP problem [28]. That is, when $\phi(S,A)$ represents the stationary probability of state S and action A, the solution of the LP problem $\phi^*(S,A)$ can be mapped to that of the CMDP-based optimization model. The equivalent LP model can be expressed as

$$\max_{\phi(S,A)} \sum_S \sum_A \phi(S,A) r(S,A), \tag{51}$$

$$\text{s.t.} \quad \sum_S \sum_A \phi(S,A) c_i(S,A) \leq \theta_i^T, \tag{52}$$

$$\sum_S \sum_A \phi(S,A) c_j(S,A) \leq \theta_j^T, \tag{53}$$

$$\sum_A \phi(S',A) = \sum_S \sum_A \phi(S,A) P[S'|S,A], \tag{54}$$

$$\sum_S \sum_A \phi(S,A) = 1, \tag{55}$$

$$\phi(S, A) \geq 0. \tag{56}$$

The objective function in (51) is to minimize the energy outage probability of IoT devices. Meanwhile, the constraints in (52) and (53) are to maintain the timer expiration probabilities of IoT devices i and j below θ_i^T and θ_i^T, respectively. The constraint in (54) satisfies the Chapman–Kolmogorov equation. The constraints in (55) and (56) are for the probability properties.

The optimal policy $\pi^*(S, A)$, which is the probability of taking a particular action at a certain state, can be obtained from the solution of the above LP problem. The optimal policy can be derived from

$$\pi^*(S, A) = \frac{\phi^*(S, A)}{\sum\limits_{A'} \phi^*(S, A')} \text{ for } S \in \mathbf{S}, \ \sum_{A'} \phi^*(S, A') > 0. \tag{57}$$

Note that, if $\sum\limits_{A'} \phi^*(S, A') = 0$, which means that there is no solution to satisfy all constraints, IoT devices do not offload any task. LP problem can be solved in polynomial time [35–37]. Therefore, our proposed algorithm can be implemented to real systems without high computational power.

4.7. IoT Device Pairing Problem

The optimal offloading policy in the previous subsection is obtained given the paired IoT devices i and j. In this subsection, we formulate an IoT device pairing problem whose objective is to minimize the summation of energy outage probabilities of all IoT devices.

Let ζ_{ij}^E denote an individual energy outage probability of IoT device i when it is paired with IoT device j. ζ_{ij}^E can be calculated as (We assume that the individual energy outage probability of IoT device i when it is paired with IoT device j is lower than that when it is not paired (i.e., $\zeta_{ij}^E < E_{ii}^E$). This assumption is reasonable because paired IoT devices operate by following the optimal policy obtained by CMDP:

$$\zeta_{ij}^E = \sum_S \sum_A \pi^* \left(\left[T_i^M, T_i^O, E_i = 0, D_i, S_j \right], A \right). \tag{58}$$

Then, the optimization problem for pairing IoT devices can be defined as

$$\min_{x_{ij}} \sum_i \zeta_{ij}^E x_{ij}, \tag{59}$$

$$\text{s.t.} \ \sum_j x_{ij} = 1, \ \forall i, \tag{60}$$

where x_{ij} is a decision variable that is 1 if IoT devices i and j are paired and 0 otherwise. The objective function in (59) is to minimize the summation of energy outage probabilities of all IoT devices. Meanwhile, the constraint in (60) to ensure that all IoT devices are paired with only one IoT device. This optimization problem is solved in the controller by using several algorithms (e.g., brute-force approach, LP relaxation, and branch-and-bound algorithm), and therefore there is no burden in IoT devices.

5. Evaluation Results

For performance evaluation, we compare the proposed algorithm, EE-CCA, with the following four schemes: (1) ALL where IoT devices always offload all of the tasks; (2) HALF where IoT devices always offload half of the task; (3) NON where IoT devices do not offload any task (i.e., process their tasks by themselves); and (4) RAND where IoT devices randomly offload their tasks at each decision epoch. For pair comparison, IoT devices are paired based on the solution of the optimization problem in Section IV-G. Meanwhile, the objective of this paper is to minimize the energy outage probability while maintaining the probability that the task is completed before their deadline above a certain level. Therefore, the average energy outage probability, ζ^E, and the average probabilities that the task of IoT device i and j are completed before their deadline, η_i and η_j, are used as performance measures of the EE-CCA. Note that η_i and η_j can be calculated by $1 - \xi_i^T$ and $1 - \xi_j^T$, respectively.

To improve the reliability of the simulation results, we have conducted over 10,000 simulation runs with different seed values independently. The default number of IoT devices is set to 6. The other default parameter settings are summarized in Table 2, where $[a\ b]$ denotes a random value between a and b.

Table 2. Default parameter settings.

τ	λ	$\mu^{F,S}$	$\mu^{F,D}$	$\mu^{H,S}$	$\mu^{H,D}$	p^H	κ	θ^T
1	[0.2 0.3]	[0.3 0.4]	[0.15 0.2]	[0.6 0.8]	[0.3 0.4]	[0.4 0.9]	[0.2 0.3]	0.99

5.1. Effect of the Harvesting Probability

Figure 4 shows the effect of the harvesting probability p_j^H of IoT device j on the average energy outage probability and the average probabilities that tasks of IoT devices i and j are completed before their deadlines. As shown in Figure 4, it can be found that the EE-CCA can reduce significantly the energy outage probability of IoT devices (see Figure 4a). For example, when P_i^H is 0.1, EE-CCA can reduce the energy outage probability by 78% compared to RAND. while maintaining the probability that the task is completed before the deadline (i.e., 0.99) (see Figure 4b,c). This is because IoT devices in the EE-CCA decide whether to offload some parts of the task to the opponent with the consideration of the energy harvesting probability, the task occurrence rate, and the current energy levels of IoT devices. For example, IoT device does not offload its task to the partner when the current energy level of the partner is low and predicted to be decreased due to low harvesting probability and high task occurrence rate.

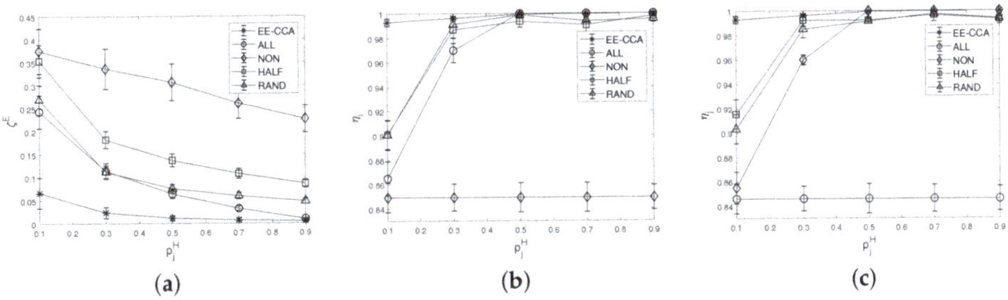

Figure 4. Effect of the harvesting probability. (**a**) average energy outage probability; (**b**) average probability that the task of IoT device i is completed before the deadline; (**c**) average probability that the task of IoT device j is completed before the deadline.

From Figure 4a, it can be shown that the average energy outage probabilities of all schemes decrease as p_j^H increases. This is because a high battery level of IoT device j can be maintained regardless of whether to offload or not when p_j^H is high. Meanwhile, from Figure 4b,c, it can be seen that the probabilities that tasks of IoT devices i and j are completed before their deadlines increases with the increase of p_H except specific cases (i.e., NON in Figure 4b and ALL in Figure 4c). This is because the tasks can be completed only when IoT devices have sufficient energy. In other words, IoT devices cannot complete their tasks within the deadline if they cannot harvest sufficiently energy. On the other hand, in NON, IoT device i does not offload its task to IoT device j. Therefore, the harvesting probability p_j^H of IoT device j does not affect the probability η_i that the task of IoT device i is completed within the deadline. Similarly, in ALL, all tasks of IoT device j are processed in IoT device i, and thus η_j is not affected by p_j^H.

5.2. Effect of the Inter-Task Occurrence Rate

Figure 5 shows the effect of the inter-task occurrence rate λ_i of IoT device i. From Figure 5, it can be found that the average energy outage probabilities of all schemes increase with the increase of the inter-task occurrence rate. This is because IoT devices consume more energy when tasks occur frequently. However, the incremental ratio of the EE-CCA is smallest among comparison schemes. This is because IoT devices in the EE-CCA operate adaptively even when the operating environment changes. Specifically, as the inter-task occurrence rate of IoT device i increases, it offloads more tasks to its opponent to avoid the energy depletion.

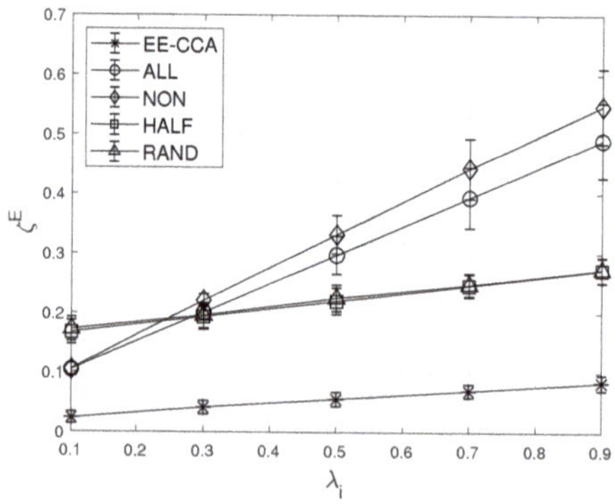

Figure 5. Effect of the inter-task occurrence rate on the average energy outage probability.

5.3. Effect of the Average Deadline

The effects of the average deadline of the task on the average energy outage probability are demonstrated in Figure 6. From Figure 6, it can be observed that the average energy outage probability of the EE-CCA decreases with the increase of the deadline. This is because, when a sufficient deadline is given, IoT devices in the EE-CCA can handle the task within the deadline by themselves even though they do not offload any tasks to energy-scarce opponents. On the other hand, the other schemes follow the fixed policy regardless of the deadline of the task, and thus their average energy outage probabilities do not change according to the deadline.

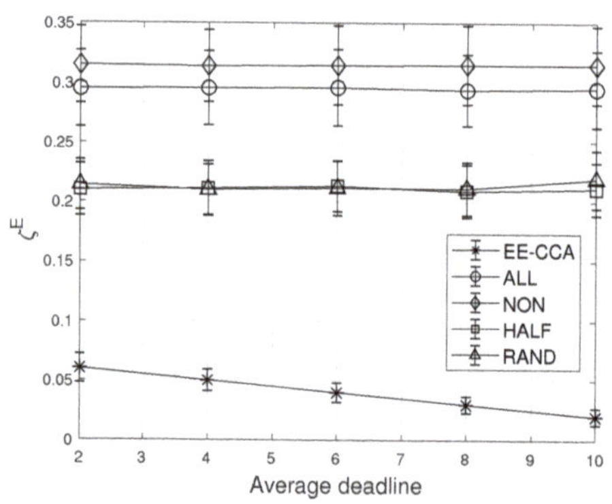

Figure 6. Effect of the average deadline of the task.

5.4. Comparison between the Optimal IoT Device Pairing and the Random Pairing

Figure 7 shows the average energy outage probabilities of the EE-CCA when pairing IoT devices based on the optimization problem (denoted by OPTIMAL) and pairing IoT devices randomly (denoted by RAND) as a function of the number of IoT devices. As shown in Figure 7, the average energy outage probability of OPTIMAL decreases as the number of IoT devices increases. This can be explained as follows: a large number of IoT devices means that there are lots of candidate IoT devices to be matched to a specific IoT device. In this situation, each IoT device can be paired to more appropriate IoT device. For example, an energy-scarce IoT device can be paired to more energy-abundant IoT device. On the other hand, since IoT devices are paired randomly in RAND regardless of the number of IoT devices, its energy outage probability is not affected by that number.

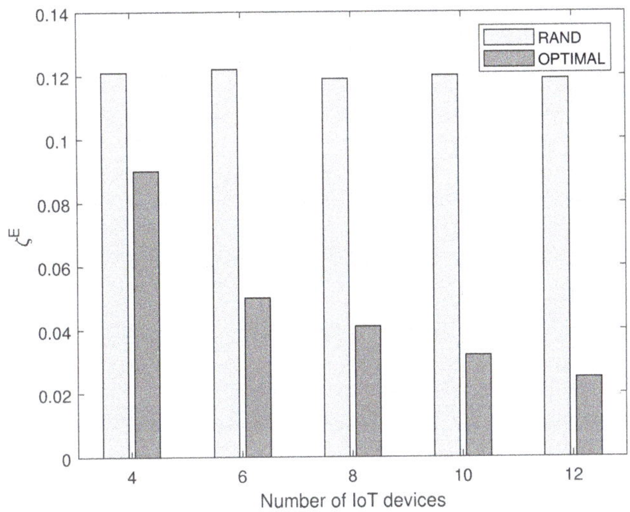

Figure 7. Comparison between the optimal pairing and random pairing.

6. Conclusions

In this paper, we proposed an energy efficient cooperative computation algorithm (EE-CCA), in which a pair of IoT devices decide whether to offload some parts of the task to the opponent with the consideration of their energy harvesting probabilities, task occurrence rates, and current energy levels. The optimal offloading decision can be obtained by means of a constraint Markov decision process (CMDP). Moreover, an optimization problem for IoT device pairing is formulated under the optimal offloading strategy. The evaluation results demonstrate that the EE-CCA offloads tasks appropriately, and thus the energy outage probability can be reduced by up to 78% compared to the random offloading scheme while providing the desired probability that tasks are completed before the deadline. Moreover, it can be seen that the EE-CCA operates adaptively even when the operating environment (e.g., inter-task occurrence rate) is changed. In our future works, we will investigate an incentive mechanism to encourage IoT devices to process tasks. In addition, a study for the robustness of the proposed algorithm will be conducted for supporting heterogeneous functionality of IoT devices.

Author Contributions: Conceptualization, H.K. and S.P.; methodology, S.J. and J.K.; software, J.L.; validation, H.K., S.J., J.K., and J.L.; investigation, S.P.; writing, original draft preparation, H.K.; review and editing, S.J., J.K., J.L., and S.P.; visualization, H.K.; supervision, S.P.

Funding: This research was supported by the National Research Foundation (NRF) of Korea Grant funded by the Korean Government (MSIP) (No. 2019R1C1C1004352).

Acknowledgments: The authors are grateful to the anonymous reviewers for their comments and valuable suggestions.

Conflicts of Interest: The authors declare no conflict of interest.

References

1. Arslan, M.; Singh, I.; Singh, S.; Madhyastha, H.; Sundaresan, K.; Krishnamurthy, S. Computing While Charging: Building a Distributed Computing Infrastructure Using Smartphones. In Proceedings of the 2012 International Conference on Emerging Network Experiment and Technology (CoNEXT), Nice, France, 10–12 December 2012; pp. 193–204.
2. Saleem, U.; Jangsher, S.; Qureshi, H.; Hassan, S. Joint Subcarrier and Power Allocation in Energy Harvesting-Aided D2D Communication Sign in or Purchase. *IEEE Tran. Ind. Inf.* **2018**, *14*, 2608–2617. [CrossRef]
3. Huang, L.; Qian, L.; Bi, S.; Xia, Z. Adaptive Scheduling in Energy Harvesting Sensor Networks for Green Cities. *IEEE Tran. Ind. Inf.* **2018**, *14*, 1575–1584. [CrossRef]
4. Bi, S.; Zeng, Y.; Zhang, R. Wireless Powered Communication Networks: An Overview. *IEEE Wirel. Commun.* **2016**, *23*, 10–18. [CrossRef]
5. Pu, L.; Chen, X.; Xu, J.; Fu, X. D2D Fogging: An Energy-Efficient and Incentive-Aware Task Offloading Framework via Network-Assisted D2D Collaboration. *IEEE J. Sel. Areas Commun.* **2016**, *34*, 3887–3901. [CrossRef]
6. He, Y.; Ren, J.; Yu, G.; Cai, Y. Joint Computation Offloading and Resource Allocation in D2D Enabled MEC Network. In Proceedings of the 2019 IEEE International Conference on Communications (ICC), Shanghai, China, 20–24 May 2019; pp. 1–6.
7. Jin, S.; Zhu, Z.; Yang, Y.; Zhou, M.; Luo, X. Alternate Distributed Allocation of Time Reuse Patterns in Fog-enabled Cooperative D2D Networks. In Proceedings of the 2017 IEEE Fog World Congress (FWC), Santa Clara, CA, USA, 30 October–1 November 2017; pp. 1–6.
8. Das, S.; Misra, S.; Wolfinger, B.; Obaidat, M. Temporal-Correlation-Aware Dynamic Self-Management of Wireless Sensor Networks. *IEEE Trans. Ind. Inf.* **2016**, *12*, 2127–2138. [CrossRef]
9. Du, R.; Gkatzikis, L.; Fischione, C.; Xiao, M. On maximizing sensor network lifetime by energy- balancing. *IEEE Trans. Control Netw. Syst.* **2018**, *5*, 1206–1218. [CrossRef]
10. Lin, C.; Deng, D.; Shu, L.; Wang, K.; Wang, S.; Tsai, I. On lifetime enhancement of dynamic wireless sensor networks with energy-harvesting sensors. In Proceedings of the 2016 EEE/CIC International Conference on Communications in China (ICCC), Paris, France, 27 June–1 July; pp. 1206–1218.
11. Wang, K.; Shen, M.; Cho, J.; Banerjee, A.; Merwe, J.; Webb, K. MobiScud: A Fast Moving Personal Cloud in the Mobile Network. In Proceedings of the 5th Workshop on All Things Cellular: Operations, Applications and Challenges, London, UK, 17–21 August 2015; pp. 19–24.
12. Tong, L.; Li, Y.; Gao, W. A Hierarchical Edge Cloud Architecture for Mobile Computing. In Proceeding of the 2016 IEEE International Conference on Computer Communications (ICC), San Francisco, CA, USA, 10–15 April 2016; pp. 1–9.
13. Taleb, T.; Ksentini, A. Follow Me Cloud: Interworking Federated Clouds and Distributed Mobile Networks. *IEEE Netw.* **2013**, *27*, 12–19. [CrossRef]
14. Liu, J.; Zhao, T.; Zhou, S.; Cheng, Y.; Niu, Z. CONCERT: A Cloud-Based Architecture for Next-Generation Cellular Systems. *IEEE Wirel. Commun.* **2014**, *21*, 14–22.
15. Puente, M.; Becvar, Z.; Rohlik, M.; Lobillo, F.; Strinati, E. A Seamless Integration of Computationally-Enhanced Base Stations into Mobile Networks towards 5G. In Proceedings of the 2015 IEEE 81st Vehicular Technology Conference (VTC Spring), Glasgow, Scotland, 11–14 May 2015; pp. 1–5.
16. Shukla, R.; Munir, A. An Efficient Computation Offloading Architecture for the Internet of Things (IoT) Devices. In Proceedings of the The 14th Annual IEEE Consumer Communications & Networking Conference (CCNC), Las Vegas, NV, USA, 8–11 January 2017; pp. 728–731.
17. Ko, H.; Lee, J.; Pack, S. Spatial and Temporal Computation Offloading Decision Algorithm in Edge Cloud-Enabled Heterogeneous Networks. *IEEE Access* **2018**, *6*, 18920–18932. [CrossRef]

18. Zhao, T.; Zhou, S.; Guo, X.; Niu, Z. Tasks Scheduling and Resource Allocation in Heterogeneous Cloud for Delay-Bounded Mobile Edge Computing. In Proceedings of the 2017 IEEE International Conference on Communications (ICC), Paris, France, 21–25 May 2017; pp. 1–7.
19. Tang, L.; Chen, X. An Efficient Social-Aware Computation Offloading Algorithm in Cloudlet System. In Proceeding of the 2016 IEEE Global Communications Conference (GLOBECOM), Washington, DC, USA, 4–8 December 2016; pp. 1–6.
20. Chen, X.; Jiao, L.; Li, W.; Fu, X. Efficient Multi-User Computation Offloading for Mobile-Edge Cloud Computing. *IEEE/ACM Trans. Netw.* **2016**, *24*, 2795–2808. [CrossRef]
21. Zheng, J.; Cai, Y.; Wu, Y.; Shen, X. Dynamic Computation Offloading for Mobile Cloud Computing: A Stochastic Game-Theoretic Approach. *IEEE Trans. Mob. Comput.* **2019**, *18*, 771–786. [CrossRef]
22. Yu, S.; Langar, R.; Fu, X.; Wang, L.; Han, Z. Computation Offloading With Data Caching Enhancement for Mobile Edge Computing. *IEEE Trans. Veh. Technol.* **2018**, *67*, 11098–11112. [CrossRef]
23. Shang, Y. Resilient Multiscale Coordination Control against Adversarial Nodes. *Energies* **2018**, *11*, 1–17. [CrossRef]
24. Zhang, W.; Wen, Y.; Guan, K.; Kilper, D.; Luo, H.; Wu, D. Energy-Optimal Mobile Cloud Computing under Stochastic Wireless Channel. *IEEE Trans. Wirel. Commun.* **2013**, *12*, 4569–4581. [CrossRef]
25. Samteladze, N.; Christensen, K. DELTA: Delta Encoding for Less Traffic for Apps. In Proceedings of the 37th Annual IEEE Conference on Local Computer Networks, Clearwater, FL, USA, 22–25 October 2012; pp. 212–215.
26. Alsheikh, N.; Hoang, D.; Niyato, D.; Tan, H.; Lin, S. Markov Decision Processes with Applications in Wireless Sensor Networks: A Survey. *IEEE Comm. Surv. Tutor.* **2015**, *17*, 1239–1267. [CrossRef]
27. Ko, H.; Pack, S. A Software-Defined Surveillance System with Energy Harvesting: Design and Performance Optimization. *IEEE Internet Things J.* **2018**, *5*, 1361–1369. [CrossRef]
28. Puterman, M. *Markov Decision Processes: Discrete Stochastic Dynamic Programming*; John Wiley and Sons: Hoboken, NJ, USA, 2009.
29. Shang, Y. Vulnerability of networks: Fractional percolation on random graphs. *Phys. Rev. E* **2014**, *89*, 1–4. [CrossRef]
30. Ko, H.; Lee, J.; Jang, S.; Kim, J.; Pack, S. Energy Efficient Cooperative Computation Algorithm in Energy Harvesting Internet of Things. Available online: https://hoy.kr/B2KEm (accessed on 10 October 2019).
31. Guo, T.; Wang, N.; Tafazolli, R. Local Mobility Management for Networked Femtocells Based on X2 Traffic Forwarding. *IEEE Trans. Veh. Technol.* **2013**, *62*, 326–340. [CrossRef]
32. Zheng, J.; Cai, Y.; Shen, X.; Zheng, Z.; Yang, W. Green Energy Optimization in Energy Harvesting Wireless Sensor Networks. *IEEE Commun. Mag.* **2015**, *53*, 150–157. [CrossRef]
33. Fu, H.; Lin, P.; Lin, Y. Reducing Signaling Overhead for Femtocell/Macrocell Networks. *IEEE Trans. Mob. Comput.* **2013**, *12*, 1587–1597. [CrossRef]
34. Ko, H.; Lee, J.; Pack, S. Performance Optimization of Delayed WiFi Offloading in Heterogeneous Networks. *IEEE Trans. Veh. Technol.* **2017**, *66*, 9436–9447. [CrossRef]
35. Spielman, D.; Teng, S. Smoothed Analysis of Algorithms: Why the Simplex Algorithm Usually Takes Polynomial Time. *IEEE J. ACM* **2004**, *51*, 385–463. [CrossRef]
36. Altman, E. *Constrained Markov Decision Processes*; Chapman & Hall: London, UK, 1999.
37. Heyman, D.; Sobel, M. *Stochastic Models in Operations Research: Stochastic Optimization*; Courier Corporation: Chelmsford, MA, USA, 2003.

energies

MDPI

Article

Two-Stage Computation Offloading Scheduling Algorithm for Energy-Harvesting Mobile Edge Computing

Laihyuk Park [1], Cheol Lee [2], Woongsoo Na [3], Sungyun Choi [4,*] and Sungrae Cho [2,*]

1 Department of Computer Science and Engineering, Seoul National University of Science and Technology, Seoul 01811, Korea; lhpark@seoultech.ac.kr
2 School of Computer Science and Engineering, Chung-Ang University, 221 Heukseok, Dongjak, Seoul 156-756, Korea; clee@uclab.re.kr
3 Media Intellectualization Research Section, Electronics and Telecommunications Research Institute, 218 Gajeong-ro, Yuseong-gu, Daejeon 34129, Korea; wsna@etri.re.kr
4 School of Electrical Engineering, Korea University, 145 Anam-ro, Seongbuk-gu, Seoul 02841, Korea
* Correspondence: sungyun@korea.ac.kr (S.C.); srcho@cau.ac.kr (S.C.)

Received: 6 September 2019; Accepted: 12 November 2019; Published: 15 November 2019

Abstract: Recently, mobile edge computing (MEC) technology was developed to mitigate the overload problem in networks and cloud systems. An MEC system computes the offloading computation tasks from resource-constrained Internet of Things (IoT) devices. In addition, several convergence technologies with renewable energy resources (RERs) such as photovoltaics have been proposed to improve the survivability of IoT systems. This paper proposes an MEC integrated with RER system, which is referred to as energy-harvesting (EH) MEC. Since the energy supply of RERs is unstable due to various reasons, EH MEC needs to consider the state-of-charge (SoC) of the battery to ensure system stability. Therefore, in this paper, we propose an offloading scheduling algorithm considering the battery of EH MEC as well as the service quality of experience (QoE). The proposed scheduling algorithm consists of a two-stage operation, where the first stage consists of admission control of the offloading requests and the second stage consists of computation frequency scheduling of the MEC server. For the first stage, a non-convex optimization problem is designed considering the computation capability, SoC, and request deadline. To solve the non-convex problem, a greedy algorithm is proposed to obtain approximate optimal solutions. In the second stage, based on Lyapunov optimization, a low-complexity algorithm is proposed, which considers both the workload queue and battery stability. In addition, performance evaluations of the proposed algorithm were conducted via simulation. However, this paper has a limitation in terms of verifying in a real-world scenario.

Keywords: computation offloading; mobile edge computing; energy harvesting; lyapunov optimization

---

1. Introduction

In recent years, along with the development of the Internet of Things (IoT) technology, it has become easier to connect mobile devices to the Internet [1–3]. In particular, IoT-based sensor devices, which have low computing performance, can overcome their computational limitations with the help of cloud systems [4,5]. However, the explosive growth of IoT data has resulted in increased traffic load in networks and cloud systems. This overload reduces the quality of experience (QoE) of the services and can result in network blackout, which shuts down the network system [6,7]. To solve these problems, the mobile edge computing (MEC) technology, which is a type of radio access network (RAN) with cloud computing capabilities, has been developed to assist resource-constrained IoT devices [6,7]. In an MEC environment, the MEC server computes the offloaded workload from the IoT devices and charges the bill accordingly. This paper proposes an energy harvesting (EH) MEC system, which enhances the survivability of the MEC through energy harvesting if the MEC is installed in a remote area where grid power supply is not available. In the EH MEC system, the MEC server is powered by renewable energy resources (RER) such as photovoltaic or wind turbine resources. Therefore, it is possible to establish a system that deploys IoT sensor nodes and collects information in places where it is difficult to install electricity facilities such as deserts or unmanned islands. However, the EH MEC system has important challenges in terms of stability. On the one hand, since the energy supply from RERs is uncertain with respect to weather or time, the EH MEC system cannot achieve stable energy unlike conventional grid powered MEC. On the other hand, if EH MEC operation only considers maximization of the system performance, system black out will occur since the power supply of EH MEC is unstable, i.e., EH MEC has to consider battery stability [8]. Thus, EH MEC reduces the energy consumption when the amount of harvested energy is not sufficient even if the system performance is decreased.

This paper proposes an EH MEC scheduling algorithm that considers the battery stability. Figure 1 shows the proposed MEC system. As shown in the figure, the IoT devices transmit the offloading requests to the MEC system. Then, the MEC system determines the admission of requests (referred to as offloading scheduling). If the offloading request is accepted, it is executed by the MEC server. Otherwise, it is transferred to the cloud system. In addition, the MEC server determines its computation frequency based on the battery state-of-charge (SoC) (referred to as MEC scheduling). According to circuit theories, the CPU power is dominated by the dynamic power, which originates from the toggling activities of the logic gates inside the CPU [9]. Thus, in this paper, we assume that the power consumption of the computation offloading can be handled by CPU frequency scheduling such as dynamic voltage frequency scaling (DVFS) [9,10]. In the proposed system, if the SoC of the battery is sufficient, the MEC server raises the computation frequency for faster offloading service. Otherwise, the MEC server will lower the frequency to ensure system stability, i.e. to avoid blackout.

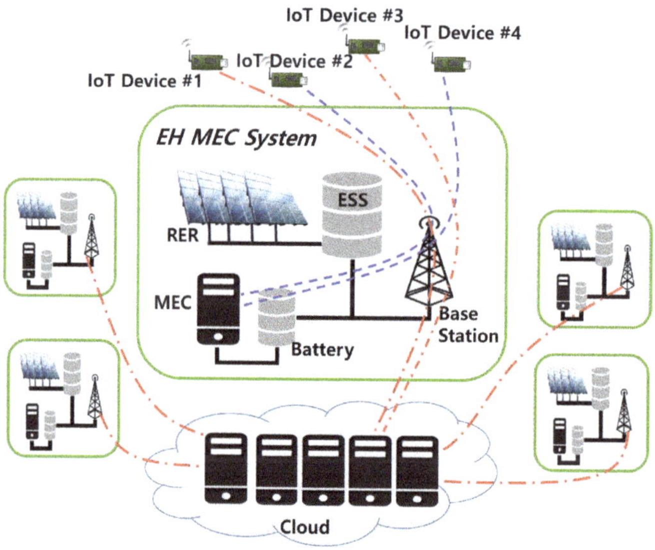

Figure 1. Architectural view of the proposed EH MEC system, consisting of RERs, energy storage system (ESS), a base station, MEC, etc. The IoT devices send their computational workloads to the base station. If the computation can be performed by the EH MEC system, it is offloaded to the MEC. Otherwise, it is offloaded to the cloud system.

The contributions of this paper are summarized as follows:

- EH MEC system: We propose an MEC integrated with RER to improve the survivability of the system through energy harvesting.The mathematical model of the MEC system includes a battery state, energy harvesting, offloading request and request deadline.
- Real-time Scheduling: We propose a two-stage scheduling algorithm, consisting of offloading and MEC scheduling. The optimization problems of the two types of scheduling are modeled as NP-hard problems. Therefore, they cannot be solved in non-deterministic polynomial time. To solve this problem, we propose a greedy algorithm to find approximate solutions for offloading scheduling. For MEC scheduling, a Lyapunov optimization-based scheduling algorithm is proposed to find the optimal computation frequency in real time.
- Battery Stability: As aforementioned, the EH MEC system has to consider the battery stability to ensure a stable operation. If the SoC is sufficient, the proposed offloading scheduler increases the number of permitted requests. Similarly, the MEC scheduler controls the frequency based on the battery SoC.

The rest of this paper is organized as follows. In Section 2, we review related works on EH MEC systems. In Section 3, the scheduling algorithms are proposed, and the corresponding optimization problem is presented. The design of the greedy algorithm, which can obtain approximate optimal solutions for offloading scheduling, and a low complexity MEC frequency scheduling algorithm based on Lyapunov optimization are provided in Section 4. Section 5 presents the performance evaluation. The conclusions of this paper are drawn in Section 6.

2. Related Work

In this section, we review the existing related works to clarify the motivation of the proposed algorithm. Over the past several years, MEC researchers have focused on maximizing computational efficiency and minimizing energy consumption of IoT devices [11–16]. In [11], the authors proposed joint offloading and computation energy efficiency maximization algorithm for MEC system. They proposed novel computation efficiency indicator and solved the problem by using iterative and gradient descent method. However, this algorithm did not consider energy harvesting. In [12], energy efficient task offloading algorithms for non-orthogonal multiple access (NOMA) MEC environment is presented. This algorithm determined the uplink power control solution, and then solved the task offloading partition and time allocation. However, this scheme also did not consider energy harvesting. The authors of [13] proposed the delay constraint offloading algorithm. This algorithm solved the problem of minimizing energy consumption, assuming that MEC server can charge IoT devices. In [14], the authors proposed a bound improving branch and bound approach to minimize energy consumption of IoT device in which orthogonal frequency division multiple access (OFDMA) was considered for uplink transmission. They focused on the energy consumption of the IoT device, under the consideration of computation offloading, subcarrier allocation, and computing resource allocation. In [15], an IoT device offloading scheduling algorithm in wireless power transfer environment is proposed. In this algorithm, the IoT device decides whether to compute the task itself or offload the task to MEC server. A cooperative partial computation offloading algorithm is proposed for MEC and cloud server environment in [16]. The authors proposed the branch and bound approach to solve the single MEC scenario, and then expanded it to multiple MEC and cloud scenario. For the multiple scenarios, an iterative heuristic MEC resource allocation algorithm was proposed. The authors of [9] proposed an algorithm for computation offloading scheduling of MEC server. They minimized the energy consumption of the MEC server as well as guaranteed the stability of the task buffer. However, it is difficult to apply this algorithm in EH environment since they did not consider energy harvesting. These previous studies [11–16] focused on the energy efficiency of IoT devices. However, this paper focuses on the energy efficiency scheduling of MEC server, i.e., MEC server has to provide stable offloading service. Due to the advancement in energy harvesting techniques, research about IoT systems with energy harvesting attracts significant attention [10,17–19]. In [17], reinforcement learning based offloading algorithm is proposed. In this algorithm, energy harvesting IoT devices decide their offloading rates according to battery levels. However, the target of this algorithm is not suitable to the proposed environment where MEC is harvesting energy. In [18], the authors proposed dynamic computation offloading algorithm with a special focus on computation capacity of MEC server. In this algorithm, the energy harvesting is not primarily considered in the offloading scheduling; it is only mentioned that it can extend network life time. A healthcare IoT system was proposed in energy harvesting environment by the authors of [19]. This algorithm is also not suitable for the EH MEC environments since it aims to protect user privacy. The authors of [10] proposed a Lyapunov optimization based algorithm for energy harvesting IoT devices. They assumed that the IoT device is equipped with an energy harvesting module and harvests electricity from the module. Each IoT devices schedules its computation frequency based on the energy harvesting status. All of the works in [10,17–19] assumed that IoT devices were equipped with energy harvesting modules and tried to solve the offloading scheduling problem under battery constraints. However, these algorithms are not suitable for the proposed EH MEC environment where the MEC server is equipped with energy harvesting module, i.e., our proposed scheme aims at the stable operation of MEC server in energy harvesting environment, but previous studies aimed at survivability of IoT devices. In [20], the authors proposed MEC scheduling algorithm in EH MEC environment. They modeled the MEC battery and harvested energy, and minimized the service delay and operation cost via reinforcement learning. The energy harvesting model is similar to the model of our proposed algorithm. However, they did not consider the deadline constraint in their problem formulation. In this paper, we consider the constraints of service deadline.

3. System Model

In this section, the mathematical modeling of the proposed EH MEC system is presented. To formulate the system model, this paper assumes that the time is divided into equal time slots indexed by t such that

$$t \in \mathcal{T} \text{ where } \mathcal{T} = \{1, 2, 3, \cdots, T\}. \tag{1}$$

This paper also assumes that each IoT device requests multiple computation offloading tasks to the EH MEC server. Therefore, the scheduling is performed for each task, and we define the notations for the set of received tasks at time t as $\mathcal{R}^t$, i.e.,

$$\mathcal{R}^t = \{r_1^t, r_2^t, r_3^t, \cdots, r_R^t\}, \tag{2}$$

where R is the number of tasks requested in the timeslot. $\mathcal{R}^t$ is stored in the request buffer in the scheduler.

As shown in Figure 1, the energy of the EH MEC system is supplied from the RER and is stored in the ESS for the operation of the MEC system. Part of the energy is transferred to the small scale battery and used to operate the MEC server. The amount of energy received by the battery is denoted as e^t, and it has a maximum value of $E_{\text{harv}}^{\max}$, i.e.,

$$0 \le e^t \le E_{\text{harv}}^{\max}, \forall t, \tag{3}$$

where the e^ts are i.i.d. in the different timeslots.

The battery level for computation offloading at the beginning of t is denoted as B^t, and it is assumed that $B^0 = 0$ and $B^t < +\infty$, i.e., B^t is consumed only for computation offloading. In addition, if we denote the energy consumed by computation offloading at t as E_{offload}^t, then the following equation holds [10]:

$$B^{t+1} = B^t - E_{\text{offload}}^t + e^t, t \in \mathcal{T}. \tag{4}$$

Based on the defined notations, the optimization problems of the computation offloading scheduling are formulated as presented in the following subsections.

3.1. Computation Offloading Model

When an IoT device requests computation offloading to the EH MEC system, the EH MEC system can either allow offloading or pass the request to the cloud server. When the EH MEC system offloads all computation tasks, the service QoE will increase and the network overhead will be reduced. However, it is impossible to offload all the computation tasks due to the constraints of the EH MEC in terms of energy or computing performance. Therefore, the scheduler of the EH MEC system controls the admission of offloading requests, i.e., it determines whether it needs to be executed by the MEC or passed to the cloud server.

Figure 2 shows the proposed EH MEC scheduler. As shown in the figure, the proposed EH MEC scheduler consists of two sub-schedulers (offloading and MEC scheduler), three buffers (request, cloud, and MEC buffers), and two managers (battery and backhaul managers). At time t, the requests from the IoT devices are stored in the request buffer, and the request buffer is modeled as $\mathcal{R}^t$. Similarly, the MEC buffer can be modeled as

$$\mathcal{M}^t = \{m_1^t, m_2^t, m_3^t, \cdots, m_M^t\}, \tag{5}$$

where M is the number of unexecuted tasks in the MEC buffer. Analogously, the cloud buffer can be modeled as

$$\mathcal{C}^t = \{c_1^t, c_2^t, c_3^t, \cdots, c_C^t\}, \tag{6}$$

where C is the number of tasks in the cloud buffer.

Based on the SoC and MEC state, the offloading scheduler performs task scheduling (i.e., admission control) and updates the MEC buffer and cloud buffer.

For offloading scheduling, this paper introduces the following mathematical models. A request for computation offloading r_i^t is defined as follows:

$$r_i^t \triangleq [L_i^t, d_i^t], \tag{7}$$

where L_i^t is the input bit size of the task and d_i^t is the execution deadline.

In this paper, we assume that d_i^t represents only the execution time in the MEC server. If we denote the number of CPU cycles required to process the bit input in the MEC server as X, then the number of CPU cycles required to execute r_i^t successfully can be obtained as

$$W_{\mathcal{R},i} = L_i^t \cdot X. \tag{8}$$

The CPU frequency of the MEC server at t and the delay for the execution of r_i^t are denoted as f^t and $D_{\mathcal{R},i}^t$, respectively. Therefore, $D_{\mathcal{R},i}^t$ can be obtained as

$$D_{\mathcal{R},i}^t = \frac{W_{\mathcal{R},i}}{f^t}. \tag{9}$$

In addition, the offloading schedule of $\mathcal{R}^t$ can be represented by

$$\mathcal{S}^t \triangleq [s_1^t, s_2^t, s_3^t, \cdots, s_R^t], \tag{10}$$

where

$$s_i^t = \begin{cases} 1, & \text{if } r_i^t \text{ is allocated to the MEC server,} \\ 0, & \text{if } r_i^t \text{ is allocated to the cloud server.} \end{cases}$$

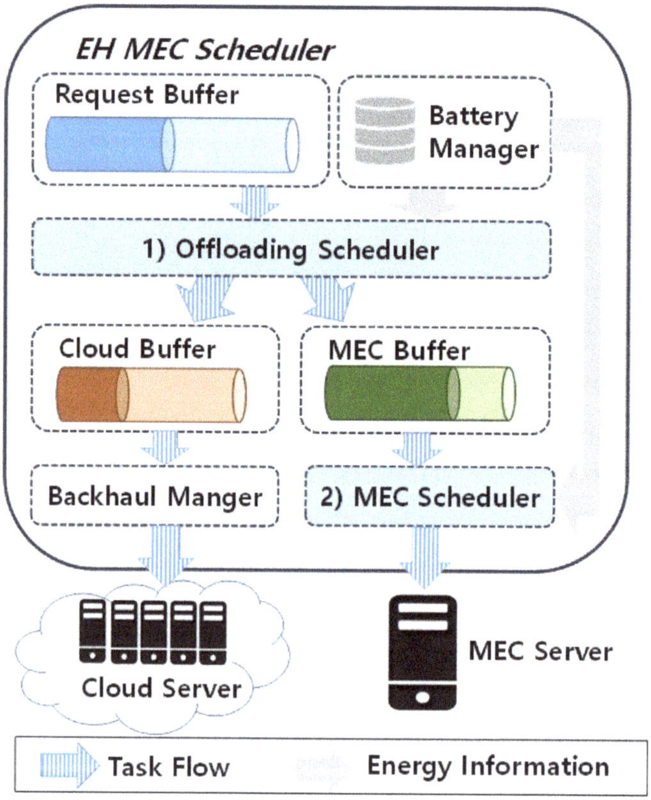

Figure 2. The proposed architecture of the EH MEC scheduler. The EH MEC scheduler includes two sub schedulers: (1) offloading scheduler; and (2) MEC scheduler.

Based on the above definitions and notations, the execution delay can be computed as follows:

$$D^t_{schedule} = \sum_{m=1}^{M} D^t_{\mathcal{M},m} + \sum_{r=1}^{R} s^t_r \cdot D^t_{\mathcal{R},r}, \tag{11}$$

where $D^t_{\mathcal{M},i}$ is the execution delay of task m^t_i in the MEC buffer and can be computed similarly to Equation (9). Therefore, the scheduling constraint for the task deadline can be derived as follows:

$$s^t_i \cdot D^t_{schedule} \leq d^t_i, \forall i. \tag{12}$$

In addition, the energy consumed for executing the task r^t_i can be obtained as follows:

$$E^t_{\mathcal{R},i} = \kappa \times W_{\mathcal{R},i} \times (f^t)^2, \tag{13}$$

where κ is the effective switched capacitance, which depends on the chip architecture. Therefore, the scheduling constraint for energy consumption can be expressed as

$$E^t_{\text{offload}} = \sum_{r=0}^{R} s^t_r \cdot E^t_{\mathcal{R},r} + \sum_{m=1}^{M} E^t_{\mathcal{M},m} \leq B^t, \tag{14}$$

where $E^t_{\mathcal{M},i}$ is the energy consumption for task m^k_i in the MEC buffer and can be computed similarly to Equation (13).

We assume that the MEC system charges the offloading bill according to the number of instruction executions, and the offloading pricing per instruction is denoted as p. Therefore, the computation offloading bill of r^t_i, denoted as $P^t_{\mathcal{R},i}$, can be obtained as

$$P_{\mathcal{R},i} = p \cdot W_{\mathcal{R},i}. \tag{15}$$

This paper also assumes that the EH MEC system wishes to maximize the bill as well as satisfy the execution delay and energy constraints. Therefore, the objective function of the offloading scheduling can be derived as follows:

$$\underset{\mathcal{S}^t}{\text{maximize}} \sum_{r=1}^{R} s^t_r \cdot P^t_{\mathcal{R},i}, \tag{16}$$

subject to

$$s^t_r \cdot D^t_{schedule} \leq d^t_r, \forall r \in \mathcal{R}, \tag{17}$$

$$\sum_{r=0}^{R} s^t_r \cdot E^t_{\mathcal{R},r} + \sum_{m=1}^{M} E^t_{\mathcal{M},m} \leq B^t. \tag{18}$$

3.2. MEC Computation Model

Based on the objective function in Equation (16) and constraints in Equations (17) and (18), the requests of the MEC buffer $\mathcal{M}^t$ can be obtained. This section describes the mathematical models used for the computation of $\mathcal{M}^t$.

To compute more bits, the MEC server increases the CPU frequency f^t according to Equation (9). However, an increase in the amount of computation results in increased energy consumption, which affects the battery life according to Equation (13).

To maximize the profit of the EH MEC system, it needs to offload more bits. However, it is impossible to maintain the maximum CPU frequency due to constraints on the battery in the EH MEC system. Therefore, the frequency scheduling of the MEC scheduler aims to maximize both the profit and battery stability.

To model the battery stability, a virtual queue Z^t is introduced. Z^t represents the amount of uncharged battery. Based on the definition of B^t, Z^t is greater than or equal to 0, i.e., $0 \leq Z^t < +\infty$. Based on Equation (4), the following equation is derived:

$$Z^{t+1} = (Z^t + E^t_{\text{offload}} - e^t)^+, \; t \in \mathcal{T}, \tag{19}$$

where

$$\alpha^+ \triangleq \max(0, \alpha). \tag{20}$$

Note that the battery should always be stable. Therefore, one constraint on the battery stability can be derived as follows:

$$\lim_{t \to \infty} \frac{1}{t} \sum_{\tau=0}^{t-1} \mathbb{E}[Z^\tau] < +\infty \tag{21}$$

The number of CPU cycles needed to execute x^t_i successfully is denoted as $W_{\mathcal{M},i}$, and it can be obtained in a similar manner as that in Equation (8). Accordingly, the required number of execution instructions at the MEC server at time t, which is denoted as I^t, can be obtained as follows:

$$I^t = \sum_{i=1}^{M} W_{\mathcal{M},i}, \tag{22}$$

where $0 \leq I^t \leq f_{\max}$.

Based on the defined notations, the number of execution cycles from the MEC buffer at time t can be obtained as

$$Q^t = \min(I^t, f^t). \tag{23}$$

A mathematical approach for maximizing the time-averaged expected number of execution cycles from the MEC buffer can be obtained as follows:

$$\text{maximize} \lim_{t \to \infty} \frac{1}{t} \sum_{\tau=0}^{t-1} Q^{\tau}. \tag{24}$$

Assuming that the computation frequency of the MEC server can be changed dynamically, the frequency scheduling vector can be defined as

$$\overrightarrow{f}^t = [f_1^t, f_2^t, \cdots, f_k^t], \tag{25}$$

where k is the number of frequency selections and f_i^t is an indicator function: if its frequency is selected, $f_i^t = 1$; otherwise, $f_i^t = 0$. Since the MEC should select only one frequency at t, the following constraints are derived:

$$\forall t \in \{0, \cdots, \infty\}, \tag{26}$$

$$\sum_{i=1}^{k} f_i^t = 1, \tag{27}$$

$$f_i^t \in \{0, 1\}. \tag{28}$$

Moreover, the selected frequency should guarantee the request deadline. Therefore, the minimum frequency scheduling vector required to guarantee all the request deadlines is denoted as k^{th}, and the following constraints can be derived:

$$\sum_{i=1}^{k^{th}-1} f_i^t = 0, \tag{29}$$

$$\sum_{i=k^{th}}^{k} f_i^t = 1, \tag{30}$$

and Equation (27) can be replaced by Equations (29) and (30).

To change $\overrightarrow{f}^t$ into a scalar value, this paper introduces the frequency selectable vector $\overrightarrow{F}$, which is defined as follows:

$$\overrightarrow{F} = [F_1, F_2, F_3, \cdots, F_k]. \tag{31}$$

Then, f^t can be calculated as

$$f^t = \overrightarrow{f}^t \cdot (\overrightarrow{F})^T. \tag{32}$$

Accordingly, Equations (20) and (24) are, respectively, modified as follows:

$$Z^{t+1} = Z^t + E^t_{\text{offload}}[f^t] - e^t, t \in \mathcal{T}, \tag{33}$$

$$\text{maximize} \lim_{t \to \infty} \frac{1}{t} \sum_{\tau=0}^{t-1} Q^{\tau}[f^{\tau}]. \tag{34}$$

Finally, the objective function, considering both the MEC buffer and battery stability, can be obtained as follows:

$$\text{maximize} \lim_{t \to \infty} \frac{1}{t} \sum_{\tau=0}^{t-1} Q^{\tau}[f^{\tau}], \tag{35}$$

subject to

$$\lim_{t\to\infty} \frac{1}{t} \sum_{\tau=0}^{t-1} \mathbb{E}[Z^{\tau}] < +\infty. \tag{36}$$

4. Proposed Scheduling Algorithm

4.1. Computation Offloading Scheduling

In Section 3.1, the mathematical models and optimization problem are designed. As shown in Equation (11), $D^t_{schedule}$ is a function of s^t_r. The constraint of the task deadline in Equation (12) is non-convex since there exists a multiplication of the function $D^t_{schedule}$ by s^t_r. Therefore, the proposed optimization problem for offloading scheduling is NP-hard and cannot be solved in non-deterministic polynomial time. To solve this problem, this paper presents a novel greedy scheduling algorithm for obtaining approximate solutions.

Algorithm 1 presents the proposed greedy algorithm for offloading scheduling. The greedy algorithm consists of offloading request sorting and offloading request scheduling. First, the sequence of offloading request scheduling based on offloading request sorting is determined. The scheduling sequence is determined based on the deadline of the requests. From Equations (11) and (12), it is obvious that a computation request having a longer deadline in the MEC buffer will be satisfied if a request with a shorter deadline is satisfied. Therefore, the proposed algorithm sorts $\mathcal{R}^t$ in the descending order of d^t_i (Line 1 in Algorithm 1). Second, the algorithm alternatively schedules each request by checking the following constraints. If r^t_i is scheduled to the MEC server, the constraints of the delay and the battery should be satisfied (Line 4 in Algorithm 1). To check the delay, the expected delay can be modified by Equation (11) and expressed as follows:

$$\widehat{D^t_{schedule}} = \sum_{m=1}^{M} D^t_{\mathcal{M},m} + D^t_{\mathcal{R},i}, \tag{37}$$

Of course, the expected delay should satisfy the execution deadline as follows:

$$\widehat{D^t_{schedule}} \leq d^t_i. \tag{38}$$

To check the battery SoC, Equation (18) can be modified as follows:

$$E^t_{\mathcal{R},i} + \sum_{m=1}^{M} E^t_{\mathcal{M},m} \leq B^t, \tag{39}$$

i.e., the sum of energy consumption for the MEC tasks and the request should not be larger than the current battery capacity.

Algorithm 1 The Greedy Algorithm.

1: **Step 1:** Offloading request sorting (descending order of d^t_i)
2: **Step 2:** Offloading request scheduling
3: **for** $i = 1 \rightarrow N$ **do**
4:     Determine if r^t_i satisfies both Equations (38) and (39).
5:     **if** r^t_i satisfies the both Equations **then**
6:         Update $\mathcal{M}^t$
7:     **else**
8:         Update $\mathcal{C}^t$
9:     **end if**
10: **end for**

If r_i^t is suitable for allocation to the MEC, it is transferred to the MEC buffer and $\mathcal{M}^t$ is updated (Lines 5 and 6 in Algorithm 1). Otherwise, r_i^t is transferred to the cloud buffer and $\mathcal{C}^t$ is updated (Lines 7 and 8 in Algorithm 1).

4.2. MEC Scheduling

In Section 3.2, the mathematical models and the corresponding optimization problem are designed. As shown in Equations (35) and (36), the proposed objective function and the constraints considered in this paper form an NP-hard problem. Our objective function is a time domain function; therefore, the Lyapunov drift optimization technique is suitable for solving this problem since we can observe the tradeoff between the performance and battery stability. Let Θ^t denote the vector of the uncharged battery queues at time t, and the quadratic Lyapunov function is defined as

$$L^t = \frac{1}{2}(\Theta^t)^T\Theta^t = \frac{1}{2}(Z^t)^2, \tag{40}$$

where $(\Theta^t)^T$ denotes the transpose of Θ^t; however, it should be noted that Θ^t has only one queue vector, and therefore Equation (40) can be derived. Let Δ^t be a conditional quadratic Lyapunov function, which can be formulated as $\mathbb{E}[L^{t+1} - L^t \mid \Theta^t]$, i.e., the drift on t [21]. The dynamic policy is designed to solve the proposed optimization formulation by observing only the current uncharged battery queue Z^t, which is maximized as follows:

$$Q^t - V\Delta^t, \tag{41}$$

where V is a positive constant value parameter used to control the drift policy, which affects the reward–battery tradeoffs [21]. Next, we select a frequency at each time slot t. By selecting a frequency, we receive a reward. This selection can be represented as follows:

$$\underset{f^t \in F_k}{\operatorname{argmax}}\; Q^t[f^t] - V \cdot Z^t \cdot (E^t_{\text{offload}}[f^t] - e^t), \tag{42}$$

where e^t is the energy harvested at t and has a constant value. Since it does not impact the results, Equation (42) can be updated as follows:

$$\underset{f^t \in F_k}{\operatorname{argmax}}\; Q^t[f^t] - V \cdot Z^t \cdot (E^t_{\text{offload}}[f^t]). \tag{43}$$

Since Equation (43) is in the closed form, the proposed algorithm can dynamically control $\overrightarrow{f}^t$ and find the optimal $\overrightarrow{f}^t$ in polynomial time.

5. Performance Evaluation

The performance of the proposed scheduling algorithms was evaluated. Through intensive MATLAB-based simulations, the following performances were verified: (1) design of the proposed algorithms; and (2) adaptation of the EH MEC environments.

5.1. Design of the Proposed Algorithm

To verify the proposed greedy algorithm, small-scale topologies ($R = 25$) were generated in a random manner. In addition, the following parameters were considered. Each request size (L_i^t) followed a uniform distribution ([5,000, 15,000] bits), and the request deadline (d_i^t) followed a uniform distribution ($[1,2]$ seconds). The battery size was assumed to be 100 Wh (=360,000 J), and the SoC of the battery followed a uniform distribution ($[30, 95]$%). The value of κ was set to 7.4×10^{-27}, and it was derived using an *Intel i7 Processor*. X was set to 740, i.e., the MEC server required 740 cycles to compute 1 bit. To approximately measure the greedy algorithm, the small-scale topologies were simulated

10,000 times with the above mentioned parameters. In the small-scale topologies, the optimal solution can be computed using the brute-force method since there are only a few offloading requests.

Figure 3 shows the difference in offloading bit size between the optimal solution and the greedy algorithm. As shown in the figure, the proposed algorithm guarantees more than 96% of the optimal value. Moreover, more than 99% of the difference between the greedy algorithm and the optimal solution are less than 700 bytes.

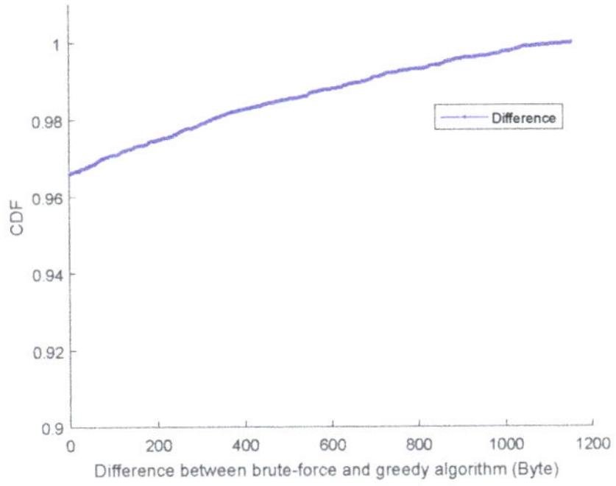

Figure 3. Difference vs. CDF.

This study also verified that the proposed MEC scheduling algorithm is well designed, i.e., the proposed Lyapunov drift optimization techniques can control the tradeoff between the performance and the battery stability according to V. For the verification of the MEC scheduling algorithms, the following simulation parameters were considered. The battery size of the MEC server was assumed to be 100 Wh (=360,000 J), and the SoC of the battery at the beginning of the simulation was assumed to be 50%.The number of IoT devices was 480, and each IoT device randomly generated the offloading request according to the Poisson distribution (the mean of the inter-arrival time was 1 s). The deadline of each request was randomly generated according to the uniform distribution using [1,2]. The number of timeslots was 86,400, and the time unit was 1 s, i.e., we simulated one day.

Figure 4 shows the changes in average battery amount and total bill according to V. As shown in the figure, the average battery amount increases with increasing V. On the other hand, the total bill decreases with increasing V. In the proposed optimization problem of MEC scheduling, there is a tradeoff between battery stability and the bill. Therefore, we can verify that the proposed Lyapunov-based MEC scheduling algorithm works well. In addition, the algorithm efficiently controls the priority between the battery stability and the bill.

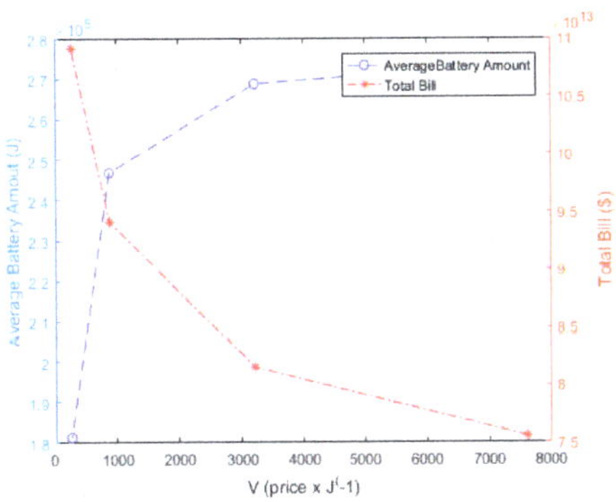

Figure 4. Average battery amount and total bill vs. V.

5.2. Adaptation of the Proposed Scheduling

This study verified the performance of the proposed scheduling algorithm in EH MEC environments. For evaluating the energy harvesting, actual measurement parameters from renewable energy resources were used. These measurement parameters were the actual field data recorded at an industrial complex that has factories making mineral waters and relevant research facilities on Jeju Island of South Korea for spring, summer, and autumn of 2017. Through the performance evaluation, we can confirm how the proposed algorithm works according to the amount of energy harvested.

Figure 5 shows the accept ratio of offloading requests according to the timeslot. In the proposed offloading scheduling algorithm, which is presented in Section 4.1, the scheduler determines the admission of requests by considering the request deadline as well as the MEC battery stability. Therefore, the proposed scheduler increases the accept ratio if the battery is sufficiently stable. As shown in the figure, the accept ratio changes according to the amount of energy harvested. At the beginning of the simulation, the requests are constantly accepted. Then, the accept ratio rapidly increases at $t = 40{,}000$ where energy harvesting begins, and then it fluctuates according to the amount of energy harvested. In addition, a higher V indicates an increasing CPU frequency; hence, the scheduler can accept more requests. Therefore, the total bill of the MEC server increases with decreasing V, as shown in Figure 4.

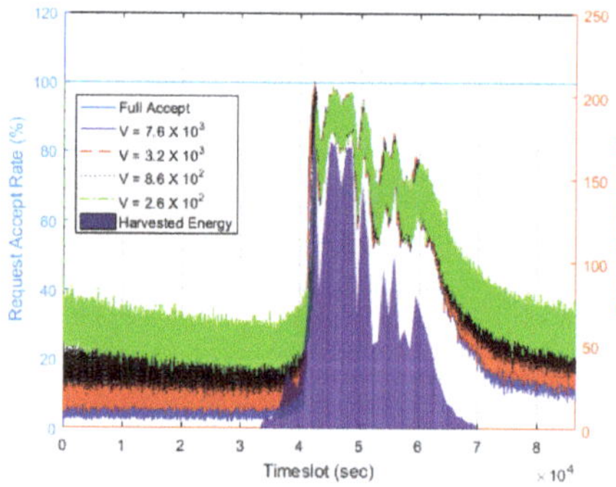

Figure 5. Request accept rate vs. timeslot.

Figure 6 shows the CPU cycles according to timeslot. As shown in the figure, the proposed scheduler efficiently controls the CPU frequency according to the amount of energy harvested. This is because the increase in energy harvesting stabilizes the battery, which results in greater battery stability. Consequently, it increases the CPU frequency to accept more requests. In addition, the figure shows the difference in number of CPU cycles with V, and a higher V relatively decreases the CPU frequency to stabilize the battery more. From $t = 0$ to $t = 40{,}000$, where the energy is not sufficiently harvested, the differences among different values of V are greater than the differences from $t = 40{,}000$ to $t = 60{,}000$, where the energy is sufficiently harvested. This is because, when the battery is not sufficiently stable, the offloading scheduler increases the number of rejected requests, and the MEC scheduler schedules the frequencies differently according to V. When the battery is sufficiently stable, however, the offloading scheduler increases the number of accepted requests and the MEC scheduler increases the CPU frequency to the maximum depending on the consumption of the battery, i.e., the CPU frequency is maximized depending on the boundaries that the battery can support.

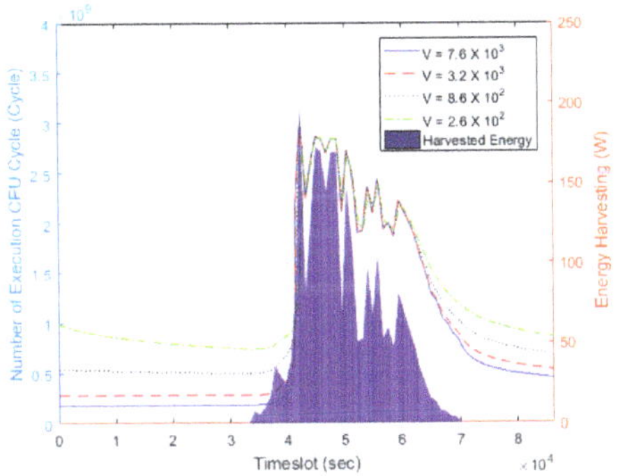

Figure 6. CPU cycles vs. timeslot.

Figure 6 shows the battery amount according to the timeslot. Similar to the previous plots, the battery amount changes according to the amount of energy harvested as shown in Figure 7. We can see that the battery amount with $V = 2.6 \times 10^2$ more rapidly compared to decreases in the battery amount with other values of V. This is because the energy consumption is quadratic with respect to the increase in CPU frequency. As shown in Figure 6, the CPU frequencies among different Vs are similar at 40,000 < t < 60,000. However, their battery amounts are different since the number of accepted requests are different.

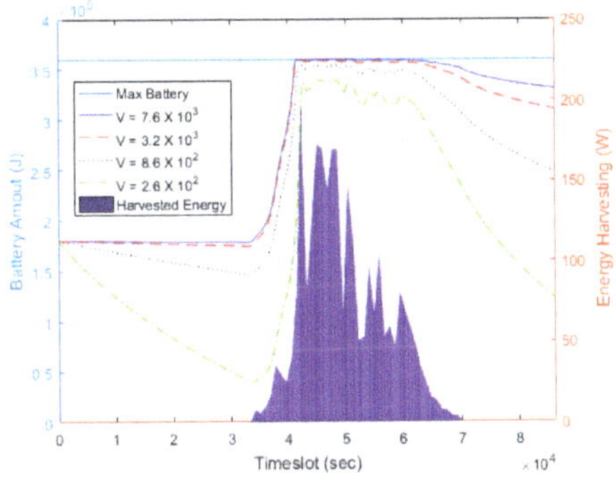

Figure 7. Battery amount vs. timeslot.

The evaluation measured the computation time of the proposed scheduling algorithm to verify its applicability. Figure 8 shows the computation time of the proposed algorithm against the number of IoT devices. As shown in the figure, the computation time is less than 0.4 ms when the number of IoT devices is 1000. In addition, the computation time increases linearly with the number of IoT devices. Thus, the complexity of the proposed scheduling algorithm is linear with the number of IoT devices. This is because the proposed scheduling algorithm is designed based on Lyapunov drift optimization, which can solve the problem in polynomial time. Therefore, the proposed algorithm can schedule the computation offloading sufficiently in real-time.

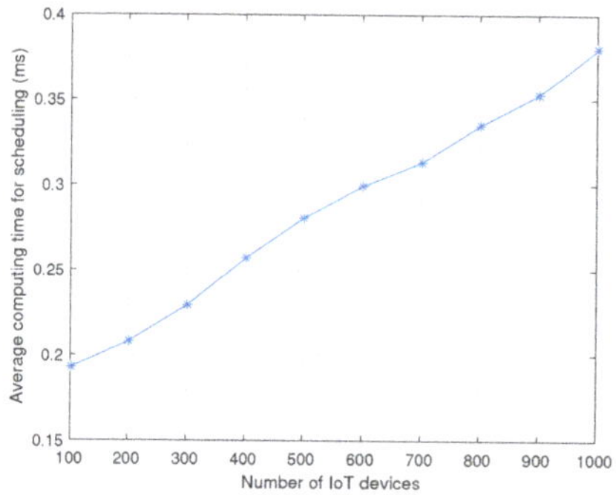

Figure 8. Computation time (ms) vs. number of IoT devices.

6. Conclusions

This paper proposes optimization formulations and their corresponding algorithms for computation offloading scheduling in an EH MEC system. For the admission control of the offloading requests, an offloading scheduling algorithm was designed to consider both the QoE of the offloading request and the system stability. The optimization problem of the offloading scheduling was formulated as a non-convex problem, and a greedy algorithm was proposed to solve the problem in polynomial time. The performance evaluation showed that the proposed greedy algorithm achieves a near-optimal performance, providing solutions that are close to the optimal solutions. To improve the survivability of the EH MEC, the proposed optimization problem of MEC scheduling considers the battery stability as well as the workload queue. Since the proposed problem of MEC scheduling is non-convex, a Lyapunov optimization-based algorithm was proposed for low-complexity scheduling. Through intensive simulations, we verified that the proposed two-stage algorithm efficiently controls the offloading requests and the CPU frequency in EH environments. However, this paper has a limitation in terms of verifying in a real-world scenario. Thus, it is essential to note that the proposed algorithm needs to be enhanced by conducting real-world testing in energy harvesting mobile edge computing. In addition, designing a scheduling algorithm by considering the offloading communication overhead is an interesting future research direction.

Author Contributions: Conceptualization, L.P.; methodology, L.P. and W.N.; software, L.P.; validation, L.P.; formal analysis, L.P.; investigation, L.P. and C.L.; resources, L.P. and S.C. (Sungyun Choi); data curation, C.L.; writing–original draft preparation, L.P.; writing–review and editing, S.C. (Sungyun Choi) and S.C. (Sungrae Cho); visualization, L.P.; supervision, S.C. (Sungrae Cho); project administration, W.N.; funding acquisition, S.C. (Sungrae Cho).

Funding: This work was supported by the National Research Foundation of Korea(NRF) grant funded by the Korea government(MSIT) (No. 2019R1F1A1064164).

Conflicts of Interest: The authors declare no conflict of interest.

References

1. Kamilaris, A.; Pitsillides, A. Mobile Phone Computing and the Internet of Things: A Survey. *IEEE Internet Things J.* **2016**, *3*, 885–898. [CrossRef]
2. Xu, L.D.; He, W.; Li, S. Internet of Things in Industries: A Survey. *IEEE Trans. Ind. Inform.* **2014**, *10*, 2233–2243. [CrossRef]
3. Perera, C.; Liu, C.H.; Jayawardena, S.; Chen, M. A Survey on Internet of Things From Industrial Market Perspective. *IEEE Access* **2015**, *2*, 1660–1679. [CrossRef]
4. Deng, S.; Huang, L.; Taheri, J.; Zomaya, A.Y. Computation Offloading for Service Workflow in Mobile Cloud Computing. *IEEE Trans. Parallel Distrib. Syst.* **2015**, *26*, 3317–3329. [CrossRef]
5. You, C.; Huang, K.; Chae, H. Energy Efficient Mobile Cloud Computing Powered by Wireless Energy Transfer. *IEEE J. Sel. Areas Commun.* **2016**, *34*, 1757–1771. [CrossRef]

6. Guo, H.; Liu, J.; Zhang, J.; Sun, W.; Kato, N. Mobile-Edge Computation Offloading for Ultra-Dense IoT Networks. *IEEE Internet Things J.* **2018**, *5*, 4977–4988. [CrossRef]
7. Fan, Q.; Ansari, N. Application Aware Workload Allocation for Edge Computing-Based IoT. *IEEE Internet Things J.* **2018**, *5*, 2146–2153. [CrossRef]
8. Zhang, D.; Chen, Z.; Ren, J.; Zhang, N.; Awad, M.K.; Zhou, H.; Shen, X.S. Energy-Harvesting-Aided Spectrum Sensing and Data Transmission in Heterogeneous Cognitive Radio Sensor Network. *IEEE Trans. Veh. Technol.* **2017**, *66*, 831–843. [CrossRef]
9. Mao, Y.; Zhang, J.; Song, S.H.; Letaief, K.B. Stochastic Joint Radio and Computational Resource Management for Multi-User Mobile-Edge Computing Systems. *IEEE Trans. Wirel. Commun.* **2017**, *16*, 5994–6009. [CrossRef]
10. Mao, Y.; Zhang, J.; Letaief, K.B. Dynamic Computation Offloading for Mobile-Edge Computing With Energy Harvesting Devices. *IEEE J. Sel. Areas Commun.* **2016**, *34*, 3590–3605. [CrossRef]
11. Sun, H.; Zhou, F.; Hu, R.Q. Joint Offloading and Computation Energy Efficiency Maximization in a Mobile Edge Computing System. *IEEE Trans. Veh. Technol.* **2019**, *68*, 3052–3055. [CrossRef]
12. Pan, Y.; Chen, M.; Yang, Z.; Huang, N.; Shikh-Bahaei, M. Energy-Efficient NOMA-Based Mobile Edge Computing Offloading. *IEEE Commun. Lett.* **2019**, *23*, 310–313. [CrossRef]
13. Feng, J.; Pei, Q.; Yu, F.R.; Chu, X.; Shang, B. Computation Offloading and Resource Allocation for Wireless Powered Mobile Edge Computing with Latency Constraint. *IEEE Wirel. Commun. Lett.* **2019**, *8*, 1320–1323. [CrossRef]
14. Yang, X.; Yu, X.; Huang, H.; Zhu, H. Energy Efficiency Based Joint Computation Offloading and Resource Allocation in Multi-Access MEC Systems. *IEEE Access* **2019**, *7*, 117054–117062. [CrossRef]
15. Wang, F.; Xu, J.; Wang, X.; Cui, S. Joint Offloading and Computing Optimization in Wireless Powered Mobile-Edge Computing Systems. *IEEE Trans. Wirel. Commun.* **2018**, *17*, 1784–1797. [CrossRef]
16. Ning, Z.; Dong, P.; Kong, X.; Xia, F. A Cooperative Partial Computation Offloading Scheme for Mobile Edge Computing Enabled Internet of Things. *IEEE Internet Things J.* **2019**, *6*, 4804–4814. [CrossRef]
17. Min, M.; Xiao, L.; Chen, Y.; Cheng, P.; Wu, D.; Zhuang, W. Learning-Based Computation Offloading for IoT Devices With Energy Harvesting. *IEEE Trans. Veh. Technol.* **2019**, *68*, 1930–1941. [CrossRef]
18. Wei, Z.; Zhao, B.; Su, J.; Lu, X. Dynamic Edge Computation Offloading for Internet of Things With Energy Harvesting: A Learning Method. *IEEE Internet Things J.* **2019**, *6*, 4436–4447. [CrossRef]
19. Min, M.; Wan, X.; Xiao, L.; Chen, Y.; Xia, M.; Wu, D.; Dai, H. Learning-Based Privacy-Aware Offloading for Healthcare IoT With Energy Harvesting. *IEEE Internet Things J.* **2019**, *6*, 4307–4316. [CrossRef]
20. Xu, J.; Chen, L.; Ren, S. Online Learning for Offloading and Autoscaling in Energy Harvesting Mobile Edge Computing. *IEEE Trans. Cogn. Commun. Netw.* **2017**, *3*, 361–373. [CrossRef]
21. Neely, M.J. *Stochastic Network Optimization with Application to Communication and Queueing Systems*; Morgan & Claypool: San Rafael, CA, USA, 2010.

energies

MDPI

Article

A Novel Coding Metasurface for Wireless Power Transfer Applications

Nguyen Minh Tran, Muhammad Miftahul Amri, Je Hyeon Park, Sa Il Hwang, Dong In Kim and Kae Won Choi *

Department of Electronic, Electrical and Computer Engineering, Sungkyunkwan University (SKKU), Suwon 16419, Korea; nmtran@skku.edu (N.M.T.); miftahulamri@g.skku.edu (M.M.A.); pjw8251@skku.edu (J.H.P.); fortyone@skku.edu (S.I.H.); dikim@skku.ac.kr (D.I.K.)

* Correspondence: kaewonchoi@skku.edu

Received: 18 October 2019; Accepted: 22 November 2019; Published: 25 November 2019

Abstract: We propose and implement a novel 1-bit coding metasurface that is capable of focusing and steering beam for enhancing power transfer efficiency of the electromagnetic (EM) wave-based wireless power transfer systems. The proposed metasurface comprises 16 $\times$ 16 unit cells which are designed with a fractal structure and the operating frequency of 5.8 GHz. One PIN diode is incorporated within each unit cell and enables two states with 180° phase change of the reflected signal at the unit cell. The two states of the unit cell correspond to the ON and OFF states of the PIN diode or "0" and "1" coding in the metasurface. By appropriately handling the ON/OFF states of the coding metasurface, we can control the reflected EM wave impinged on the metasurface. To verify the working ability of the coding metasurface, a prototype metasurface with a control board has been fabricated and measured. The results showed that the coding metasurface is capable of focusing beam to desired direction. For practical scenarios, we propose an adaptive optimal phase control scheme for focusing the beam to a mobile target. Furthermore, we prove that the proposed adaptive optimal phase control scheme outperforms the random phase control and beam synthesis schemes.

Keywords: 1-bit unit cell; coding metasurface; adaptive beam focusing; wireless power transfer

---

1. Introduction

In the last decade, due to the increasing number of wireless and mobile devices being used, charging these devices has become a crucial problem, which is now capturing massive attention. The traditional charging method with cords is not preferable for futuristic systems and devices (e.g., Internet of Things, wireless sensor networks (WSN)). Wireless power transfer (WPT), allowing us to charge a device without any wire, is emerging as a promising technology for resolving the battery charging problem.

Since the first demonstration of WPT by Tesla in early 1900, engineers and researchers have recently come up with various techniques to transfer power wirelessly [1–3]. Among them, electrical methods, consisting of inductive coupling, magnetic resonant coupling, and electromagnetic (EM) radiation, are pervasive. While the first two methods are able to provide high transmission efficiency within a short range, EM radiation is capable of providing long range WPT but with low efficiency [2]. Achieving both long range and high efficiency is still the main challenge in WPT which requires optimal solutions. For the case of EM radiation, beam forming and focusing is a potential solution for both long distance and high efficiency, which is enabled using phased-array antennas (PAA).

Recently, a novel concept of coding and digital metasurface has emerged as a promising and alternative technique that can perform beam focusing, multibeam, or scattering [4,5]. A metasurface consists of hundreds to thousands of unit cells which have different reflection responses to incident EM waves. In contrast to PAA, which depends on power-hungry and active components, coding metasurface enables beam synthesis by just turning ON or OFF the PIN diode integrated with each unit cell in the metasurface. By doing that, we can actually control the reflected signal outgoing from the metasurface.

The term coding metasurface was first coined by Cui and colleagues in 2014 [4]. Since then, several works and progress have been made in this area [5–7]. Different designs of 1-bit unit cell and metasurface have been proposed in [8–13]. Yang and colleagues proposed a simple rectangular element but working effectively in metasurface. The authors first designed and fabricated 10×10 array to demonstrate the working ability of the unit cell and metasurface [8,9]. Afterwards, this work was extended to 1600 element metasurface [10–12]. In these papers, the authors theoretically and experimentally demonstrated that the 1-bit metasurface is capable of digital beam focusing, multibeam focusing, scattering, and broadside beam synthesis. A 1-bit digital reconfigurable reflective metasurface with 20×20 cells is presented for beam-scanning in [13]. In that work, the authors used a varactor diode instead of a PIN diode for achieving wider bandwidth. Some works have also focused on designing 2-bit metasurface [14,15]. For instance, a dynamic beam manipulation based on 2-bit digitally controlled coding metasurface was proposed in [14] by Huang and colleagues. To enable a 2-bit operation, the authors used two PIN diodes in each meta-atom to produce four phase responses of 0, $\pi/2$, π, and $3\pi/2$, which correspond to four basic digital elements "00", "01", "10", and "11". Experiments and measurements were conducted, which demonstrate one-beam deflection, two- and four-beam splitting, and beam diffusion by real-time control of the bias voltage. Another work is a transmission-type 2-bit programmable metasurface for single-sensor and single-frequency microwave imaging in [15]. The authors designed the unit-cell with two layers, in which each layer contains a switchable diode for providing 2-bit control ability. Recently, there has been an increasing number of research works which focus on using metasurface in wireless power transfer [16–22]. Nevertheless, these works are just able to do fixed focusing and are definitely inapplicable for mobile devices.

For an adaptive beam focusing, the channel between each unit cell of the metasurface and the receiver should be estimated. The authors in [23–25] theoretically analyzed and proposed the method for estimating the channel in intelligent surface or programmable metasurface. The method enables channel estimation by setting a single unit cell in "ON" state and the others in "OFF" state in a training time. However, this method might be impractical, especially when it comes to huge number of unit cells in the metasurface, as the reflected signal from one unit cell is too feeble compared to the rest.

In this paper, we propose a novel 1-bit coding metasurface that can dynamically perform beam focusing to the desired direction for WPT applications. Indeed, the proposed metasurface consists of 16×16 unit cells which are designed to operate at 5.8 GHz with two states (ON/OFF states) corresponding to a 0° or 180° phase shift of the reflected signal. To obtain the unit cell for a real metasurface, we applied the fractal structure in designing the unit cell. Therefore, the unit cell has ca ompact size with dimensions of $11 \times 11 \times 1.52$ mm^3. In order to electrically steer the beam, appropriate ON/OFF states of the unit cells in the metasurface should be set by a control board. The beam-focusing ability of the proposed coding metasurface has been validated by the experiment. The experimental results showed that the metasurface can steer and focus the beam within the range of $(60^\circ, 0^\circ)$ in the elevation angle. For practical scenarios, an optimal phase control scheme is proposed and applied to adaptively track the mobile devices. The experimental results showed that the optimal phase control scheme performs better than the random phase control and beam synthesis schemes.

2. Coding Metasurface Theory

In contrast to the existing metasurfaces that change the design structure of each unit cell to acquire the desired reflective phase shift, the coding and digital metasurface manipulates and reflects the impinged EM wave via different states of identical unit cells with the help of PIN diodes. For the theoretical analysis, we considered a coding metasurface with $M \times N$ 1-bit unit cells as shown in Figure 1. Specifically, as incident with an EM wave, the unit cell can operate in two states with the same reflected magnitude but 180° phase change. There are possibly two types of illuminating sources (plane wave and a point source) to be considered.

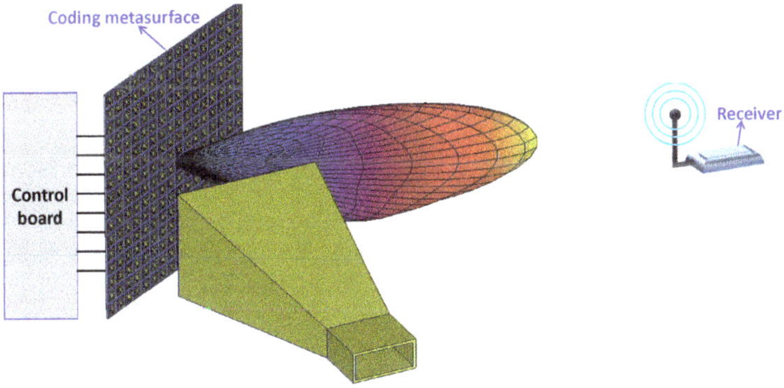

Figure 1. The coding metasurface model.

With x- or y-polarized EM wave incidence, the scattering field from the coding metasurface can be theoretically expressed as [12]

$$E(\theta,\varphi) = \sum_{m=0}^{M-1}\sum_{n=0}^{N-1} A_{mn}e^{j\alpha_{mn}} \cdot \Gamma_{mn}e^{j\phi_{mn}} \cdot f_{mn}(\theta,\varphi) \cdot e^{jk_0(md_x \sin\theta\cos\varphi + nd_y \sin\theta\cos\varphi)}, \quad (1)$$

where A_{mn}, α_{mn} are the relative illuminating amplitude and phase with respect to each unit cell in the metasurface ($A_{mn} = 1, \alpha_{mn} = 0°$ if the source is plane wave), Γ_{mn}, ϕ_{mn} are the reflection amplitude and phase of mnth unit cell, $f_{mn}(\theta,\varphi)$ is the scattering pattern of the unit cell, and d_x and d_y indicate the unit cell spacing in x and y directions.

According to Equation (1), the scattering EM wave from the metasurface can be controlled and formed by adjusting the reflection amplitude (Γ_{mn}) and phase (ϕ_{mn}) of each unit cell. Therefore, we can say that the metasurface may possibly be encoded via these two parameters. Assuming that the reflection magnitude is identical, the reflection phase matrix with respect to the coding metasurface can be described by:

$$\Phi = \begin{bmatrix} \phi_{11} & \phi_{12} & \cdots & \phi_{1n} \\ \phi_{21} & \phi_{22} & \cdots & \phi_{2n} \\ \vdots & \vdots & \ddots & \vdots \\ \phi_{m1} & \phi_{m2} & \cdots & \phi_{mn} \end{bmatrix}_{M\times N}. \quad (2)$$

One should obtain the appropriate Φ matrix for a specific beam pattern synthesis. To focus the beam to the desired direction, the phase compensation of each unit cell can be calculated by [21]:

$$\phi_{mn} = k(|\mathbf{f} - \mathbf{r}_{mn}| + |\mathbf{d} - \mathbf{r}_{mn}|), \quad (3)$$

where k is the wavenumber, $\mathbf{f}$ is the location of the EM source, $\mathbf{d}$ is the location of the focusing point, and $\mathbf{r}_{mn}$ is the position of the mnth unit cell.

2.1. Phase Quantization

The above computed phase is the ideal phase shift of each unit cell. However, only a limited phase shift can actually be provided with the coding unit cell. Hence, it is inevitable that we have to consider the phase quantization for the coding metasurface [13]. One can realize from Equation (3) that the phase compensation might be not within the range between 0 and 2π. Then, it should be shifted to be in the range of $[0, 2\pi]$ before being quantized. Considering 1-bit coding unit cell, the shifted and quantized phase can be given as:

$$\bar{\phi_{mn}} = \phi_{mn} - 2\pi \left\lfloor \frac{\phi_{mn}}{2\pi} \right\rfloor, \tag{4}$$

$$\hat{\phi_{mn}} = \begin{cases} \pi & \text{if } \pi/2 \leq \bar{\phi_{mn}} \leq 3\pi/2 \\ 0 & \text{elsewhere.} \end{cases} \tag{5}$$

After quantization, the reflection phase of the unit cell has just two states: 0° or 180°, corresponding to the coding "1" and coding "0", respectively. To demonstrate the focusing ability of the coding metasurface, we computed the ideal phase and quantized phase distribution of the 16 × 16 coding metasurface using the above phase compensation method and the corresponding beam pattern as presented in Figure 2. The metasurface is assumed to be placed in XOY plane and is excited by a point source. Although the relative power generated by the ideal phase distribution (245) is much higher compared to one of the quantized phase distribution (around 170), both cases are capable of performing good beam steering. It is clear that beam synthesis can be acquired by 1-bit coding metasurface.

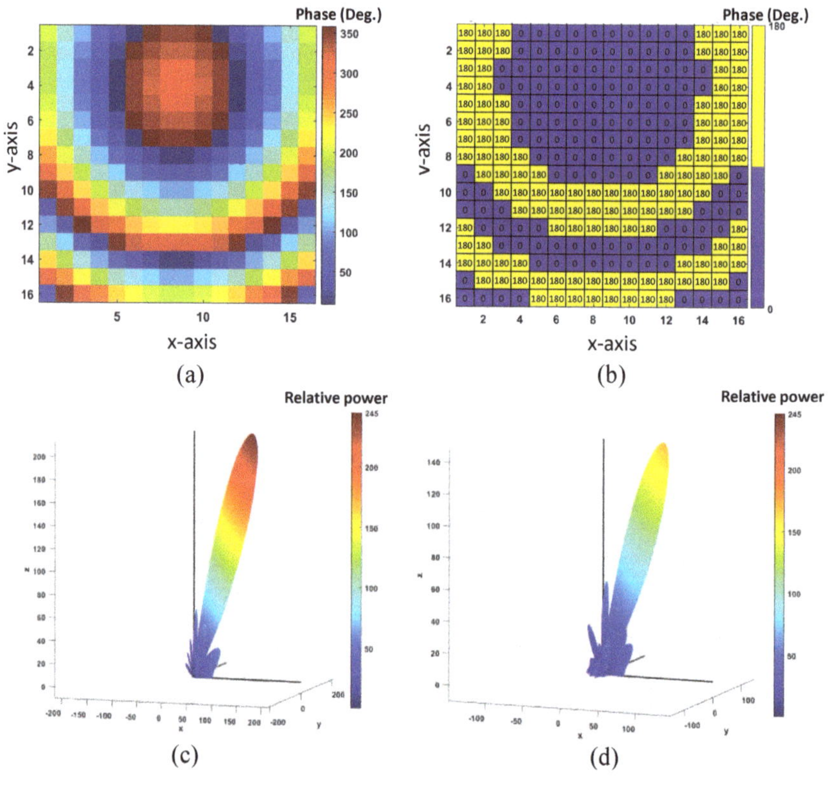

Figure 2. Reflection phase distribution and 3D pattern of the coding metasurface with an electromagnetic (EM) source located at (0 cm, 0 cm, 10 cm) and beam focusing at (0°, 30°): (**a**) ideal reflection phase distribution; (**b**) quantized phase distribution; (**c**) the beam pattern w.r.t ideal phase distribution; (**d**) the beam pattern w.r.t quantized phase distribution.

2.2. *Optimal Phase Control for Adaptive WPT*

Due to the imperfection of the manufacturing and experimental setting, the beam may not exactly be directed to the desired direction, which reduces the power received at the receiver. Moreover, in the practical scenarios, the coding metasurface should perform adaptive beam tracking according to the position of the mobile receiving devices. In order to tackle the above-mentioned problem, we propose an optimal phase control scheme allowing us to localize the receiver and focus the power toward that desired position based on the experimental data. The procedure of obtaining the adaptive optimal phase is presented in Figure 3.

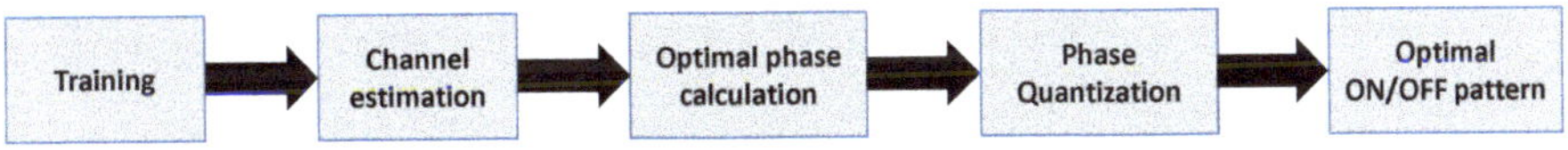

Figure 3. The block diagram of the proposed adaptive optimal phase control scheme.

Each unit cell in the metasurface can be considered as an antenna element in an array antenna. Therefore, we can mathematically express the system model as below:

$$y = \sum_{i=1}^{M} h_i x_i + n, \tag{6}$$

where M is the number of unit cells in the metasurface, n is the additive white Gaussian noise, and h_i is the channel corresponding to the ith unit cell with the state x_i.

For adaptive beam focusing, we have to know the channel to obtain the optimal phase at the metasurface. Analogous to the multiple input single output (MISO) system, we can estimate the channel from each unit cell to the receiver using predefined pilots. Recently, researchers in [23–25] attempted to estimate the channels of the intelligent surface by sending data with one unit cell in "ON" state and the others in "OFF" state at a given training time. However, this method would be impractical, especially when it comes to a huge number of unit cells in the metasurface, as the reflected signal from only one "ON" unit cell is too feeble compared to the rest. Consequently, this leads to unfeasible measurement of the change in the received signal at different training times. Therefore, to tackle this problem, we used 256 independent ON/OFF patterns of the metasurface, which is based on the Hadamard matrix, as training pilots. Then, the adaptive optimal phase control scheme is proposed with the following steps:

- **Training**: sending the signal with 256 independent ON/OFF patterns;
- **Channel estimation**: 256 channels between each unit cell of the metasurface is estimated by multiplying the received signals with the inversion of 256 transmitting patterns;
- **Optimal phase calculation**: optimal phase of each unit cell is obtained by taking the phase of the channel after being conjugated;
- **Phase quantization**: quantizing the phase based on Equation (5);
- **Optimal ON/OFF pattern**: mapping the quantization phases with the ON/OFF state (180° to "ON", 0° to "OFF").

3. Design of Coding Metasurface

3.1. *Unit Cell Design*

In order to have a good performance coding metasurface, a unit cell should be precisely designed to have particular characteristics. As the ISM (industrial, scientific, and medical) band is free and pervasive, we selected 5.8 GHz which is in the ISM band as the target operating frequency in this work. The final 1-bit unit cell structure is presented in Figure 4. The unit cell is designed on a substrate of Taconic RF-35 with a permittivity of 3.5 and loss tangent of 0.0018. A PIN diode (SMPA1320-079LF-203195B) is integrated with each unit cell to connect the main unit cell with the ground plane through a metal via hole. Thus, it can enable the two states of the unit cell. The operation of the unit cell is ensured by a bias line which is connected to the control board.

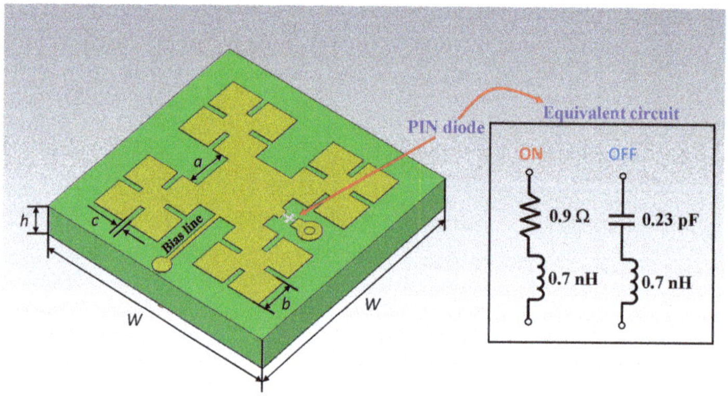

Figure 4. Unit cell structure.

The equivalent circuit for the two states of the PIN diode is given in Figure 4. As can be seen, in an ON state, the PIN diode behaves as a series circuit of a resistance and an inductance, whereas it acts as a series circuit of a capacitance and an inductance. Under an EM wave illumination, the impedance of the PIN diode can be described as:

$$Z_{pin}(\omega) = \begin{cases} R + j\omega L & \text{ON state} \\ j\omega L + \frac{1}{j\omega C} & \text{OFF state.} \end{cases} \tag{7}$$

where $R = 0.9\ \Omega$, $L = 0.7$ nH, and $C = 0.23$ pF. The EM wave will be reflected at the PIN diode and be re-radiated at unit cell surface with the reflection coefficient, which is calculated as:

$$\Gamma(\omega) = \frac{Z_{pin}(\omega) - Z_R}{Z_{pin}(\omega) + Z_R} = |\Gamma| e^{j\phi} \tag{8}$$

where Z_R is the radiation impedance of the unit cell. One should find the structure which provides an appropriate value of Z_R to obtain a 180° phase change between ON and OFF states at the objective frequency. In order to easily figure out the proper value of radiation impedance and reduce the uncontrolled reflection from the unit cell, the unit cell structure should be designed to resonate at the objective frequency. Normally, a rectangular patch with approximately half wavelength dimensions is used as a resonant structure. However, in this work, we used the fractal structure to achieve compact size (a quarter wavelength) but provided the same performance as the rectangular patch. The detailed dimensions of the unit cell are given in the Table 1.

Table 1. The unit cell dimensions.

Parameter	Value (mm)	Parameter	Value (mm)
W	11	h	1.52
a	1.83	b	1.52
c	0.25		

The unit cell is simulated using commercial software CST studio suite and the simulated reflection magnitude and phase of the unit cell are shown in Figure 5. It is obvious that at 5.8 GHz, while the reflection magnitude is almost identical between the ON and OFF state, the reflection phase between the two cases has 180° change. The maximum unit cell loss is around 0.5 dB for the OFF state. It is evident that the proposed 1-bit unit cell is suitable for the coding metasurface. As the phase change between ON/OFF states is relative, we can simply state that a unit cell with an ON state corresponds with a 180° phase reflection and one with an OFF state has a 0° phase reflection.

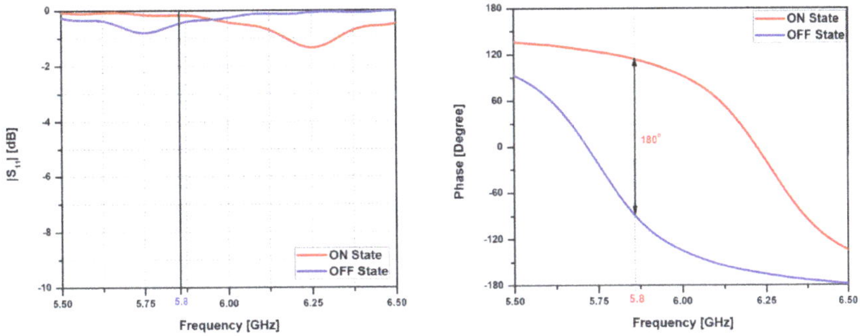

(a) (b)

Figure 5. The simulation results of the unit cell: (**a**) reflection magnitude; (**b**) reflection phase.

3.2. Coding Metasurface Construction

With the proposed unit cell, a 1-bit 16 × 16 coding metasurface is constructed and fabricated using printed circuit board technology, as shown in Figure 6. The metasurface has a total size of 176 mm × 176 mm and is precisely attached to the output pins of the control board as a sandwich structure. To enable the beam focusing ability, the control board plays a crucial role in coding metasurface, which provides an exact voltage to turn the PIN diode in each unit cell ON and OFF. The block diagram and prototype of the control board are presented in Figure 7.

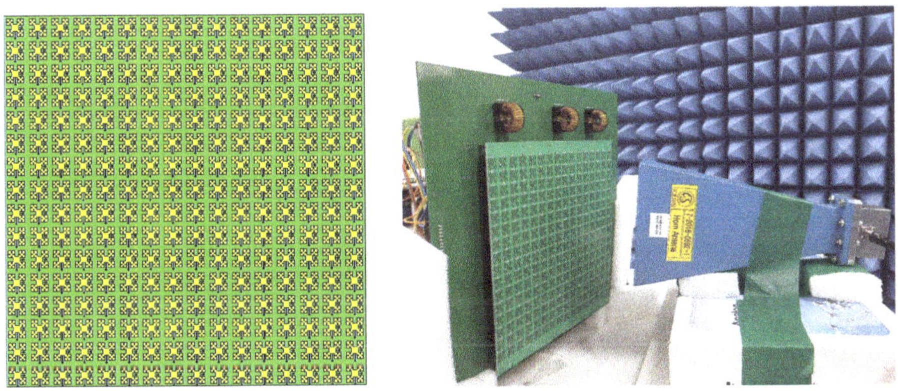

(a) (b)

Figure 6. The 1-bit 16 × 16 coding metasurface model and prototype: (**a**) metasurface model; (**b**) experimental set-up.

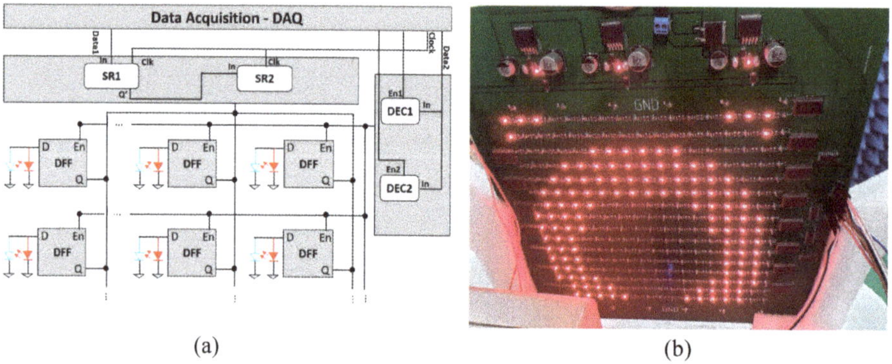

Figure 7. The control board of the coding metasurface: (**a**) block diagram; (**b**) the prototype.

As the coding metasurface has 256 unit cells, the control board has to control 256 output pins independently. In order to reduce the complexity of wiring the control board, we used two 8-bit shift registers (SR1 and SR2) to independently set the data for each row. Then, the data will be stored in the D flip flip (DFF), which is enabled by two 3 to 8 line decoders (DEC1 and DEC2). Therefore, by selecting the appropriate row, the data can be independently loaded to the intended row. Consequently, we can control the coding metasurface with any ON/OFF state of the unit cell. The LED is parallel connected to the PIN diode to indicate the state of the unit cell, which gives an observable view of the active ON/OFF pattern of the coding metasurface. All the input data are provided by the data acquisition (DAQ) that is controlled by a LabVIEW program.

4. Results

4.1. Simulation Results

Before fabricating the prototype, a 16×16 coding metasurface is modeled and simulated using CST Studio software to verify the theory. In this simulation, the metasurface is in the XOY plane, and a horn serves as an EM source, which is located at (-5.7 cm, 0 cm, 10 cm) or at $(-30^{\circ}, 0^{\circ})$ with respect to the metasurface. Then, an ON/OFF pattern matrix for steering to $(40^{\circ}, 0^{\circ})$ is loaded to the PIN diode of each unit cell. Finally, the simulation results are exported and shown in Figures 8 and 9.

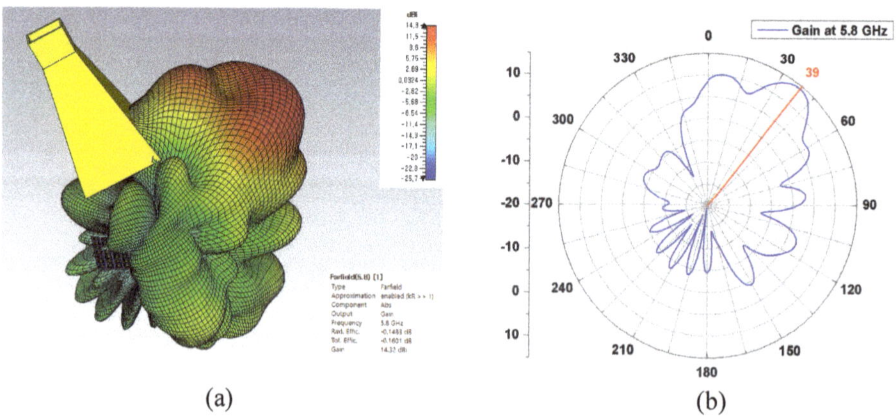

Figure 8. The simulation results: (**a**) The 3D gain total with the horn and metasurface; (**b**) the 2D radiation pattern.

It is clearly observed that the coding metasurface is capable of steering the beam to the desired direction with 1° error in the test case. Furthermore, the surface current distribution in Figure 9 is almost coincident with the input ON/OFF coding pattern. Specifically, we can notice that the current distribution of the unit cell in OFF state is considerable compared to the one with the ON state.

As the proposed system transfers the power via EM waves, human exposure and the specific absorption rate (SAR) level should be considered. Figures 10 and 11 present the simulation SAR level (averaged over 1 gram of human tissue) and beam shape when a human head is exposed closely to the metasurface. In this simulation, the metasurface is encoded to focus the beam to (30°, 0°) at the human head, which is placed 50 cm away from the metasurface. The transmitted power is 27 dBm. Figure 10 demonstrates that SAR level at the operating frequency is within the specified limit of 1.6 W/kg regarding to FCC limit. In addition, in the case of human exposure, the beam shape is a bit wider compared to the one without a human head. This results in degrading the gain from 17.3 to around 15 dBi, as shown in Figure 11.

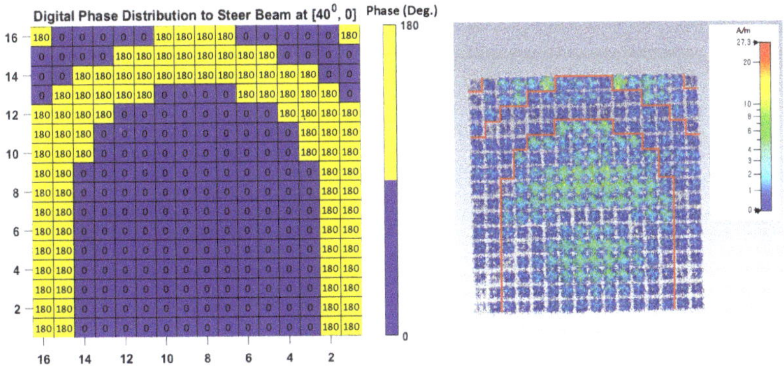

(a) (b)

Figure 9. The ON/OFF coding pattern of the coding metasurface: (**a**) calculated phase distribution matrix; (**b**) simulated current distribution.

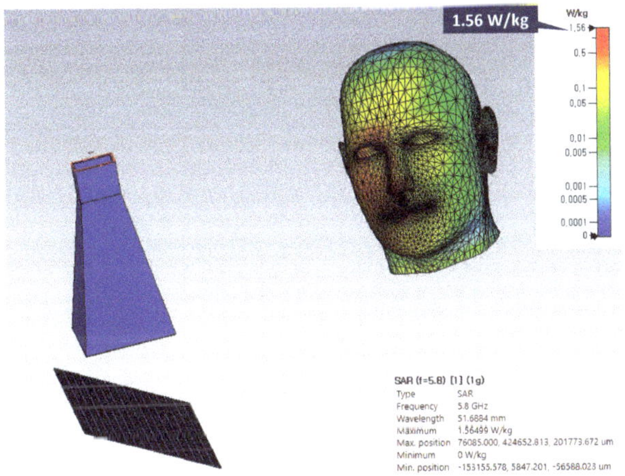

Figure 10. Simulation SAR results with a human head (50 cm away from the metasurface).

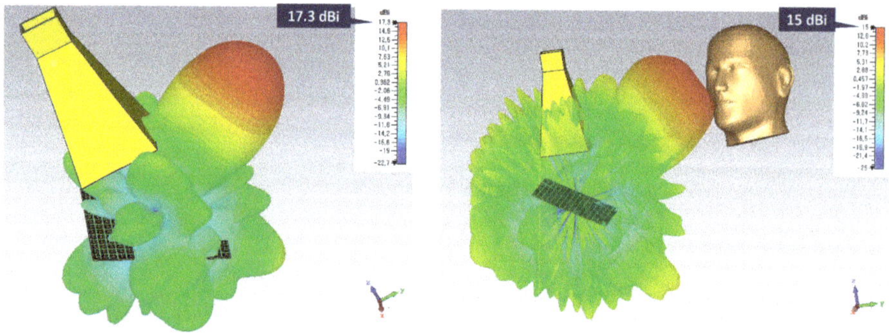

(a) (b)

Figure 11. Effect of human exposure in beam shaping: (**a**) without a human head; (**b**) with a human head.

4.2. Experimental Results

To demonstrate the capability of the proposed coding metasurface, after fabricating, several tests were done with the prototype.

4.2.1. Beam Steering with Beam Synthesis Scheme

In the first test, we applied ON/OFF coding patterns, which were calculated using the beam synthesis scheme, to verify the beam steering capability of the metasurface. The metasurface is attached to a holder which is perpendicular to the horizontal plane as shown in Figure 7b. A horn antenna working as an EM source is placed at (10 mm, −5.7 mm, 0 mm) with respect to the metasurface to avoid blocking the reflected signal. The measured results are presented in Figure 12.

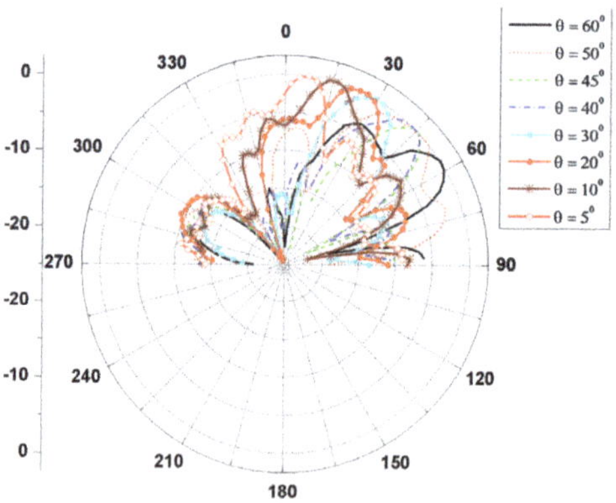

Figure 12. The measured normalized radiation pattern.

We can observe that the coding metasurface can perform beam focusing and steering to a specific direction with the elevation angle (θ) range of 0° to 60°. As a matter of fact, the steerable elevation angle can range from -60° to 60° and the azimuth angle, which the main beam can be focused on, is $(0^\circ, 360^\circ)$ by properly assembling the feed horn. However, in the limitation of this paper, we would not demonstrate these results. Further, there are constant errors of about 2° occurred in every cases that may be caused by the imperfection of the experimental setup. Moreover, similar to the theory, the grating lobe appears in the case of steering the beam to 50° and 60°. It is the limitation of the planar metasurface. In addition, the half-power beamwidth (HPBW) ranging from 13° to 27° is likely large due to the small size of the coding metasurface. A much sharper beamwidth can be attained with a larger coding metasurface.

4.2.2. Adaptive Beam Steering with Optimal Phase Control

To validate the effectiveness of the optimal phase control scheme, firstly, we present the optimal quantized phase distribution and the corresponding beam pattern which is compared to the one from the beam synthesis scheme in Figure 13. It can be observed from Figure 13b that a better beam pattern with a higher transmission coefficient is achieved with optimal phase distribution in comparison with the one of a beam synthesis scheme.

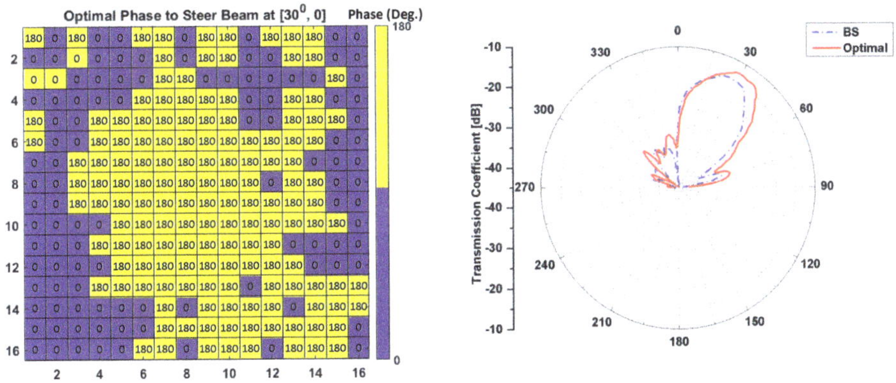

(a) (b)

Figure 13. Optimal phase control scheme results: (**a**) optimal quantized phase distribution for steering beam at (30°, 0°); (**b**) comparison of beam pattern between optimal phase control (Optimal) and the beam synthesis (BS) schemes.

Furthermore, we measured and compared the efficiency of power transfer between the optimal phase and random phase control schemes in Figure 14a. We also compared the optimal phase control and the beam synthesis schemes in Figure 14b. As can be seen from Figure 14, the proposed scheme demonstrates much higher efficiency in comparison with the random phase control scheme. At the same time, it is clear that the optimal one outperforms the beam synthesis scheme as indicated in Figure 14b. While around 4% efficiency is observed in the optimal phase control scheme at 50 cm with the steering angle of 30°, only around 3% efficiency is achieved in beam synthesis scheme. By extending the size of the coding metasurface, higher power transfer efficiency will be achieved.

The above results were acquired considering only line of sight transmission. However, in actual WPT scenarios, obstacles such as humans and animals might be exposed between the transmitter (Tx) and the receiver (Rx). Hence, this would have a severe effect on power transmission efficiency. To comprehend this problem, we investigated the power transfer efficiency as a human hand and body inserted between Tx and Rx, and the results are indicated in Figure 15. It is evident that the efficiency declines almost 1% with hand blocking and almost drops to 0% with human body blockage.

Actually, we can redo the training to get the optimal phase to enhance the efficiency when the human phantom is inserted. We did re-training with a human body exposed in some positions and compared with the results without re-training as presented in the Table 2. It is clear that remarkable improvement can be achieved by re-training the programmable metasurface when humans are exposed between Tx and Rx.

Table 3 shows the comparison of the proposed system with the previous works. With a fixed beam and large dimensions, the reflect array in [26] provides a higher efficiency compared to the phased array in [27] and our work. This results not only from the large size of the surface but also from the almost ideal phase distribution used in focusing to a fixed position in this work. This shows the potential of using a metasurface in WPT. The phased array in [27] outperforms our work in transfer distance operating in the lower frequency, which suffers lower loss from transmission path but leads to a physically large system. The performance of our proposed programmable metasurface can be enhanced with a larger size of the metasurface.

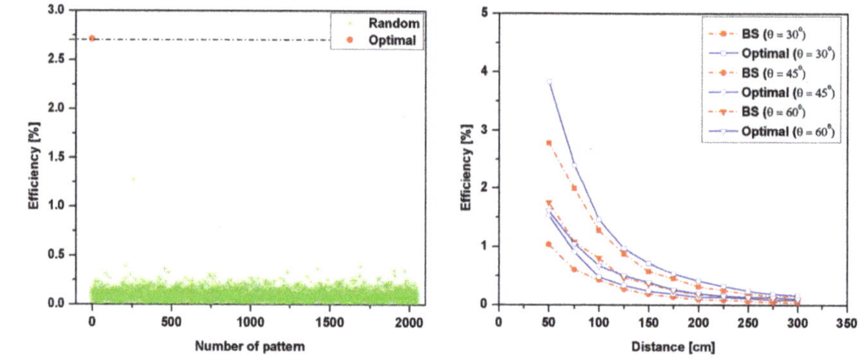

(a) (b)

Figure 14. Power transmission efficiency comparison: (**a**) optimal phase control and random phase control schemes with the steering angle at (30°, 0°); (**b**) optimal phase control (optimal) and beam synthesis (BS) schemes at different distances.

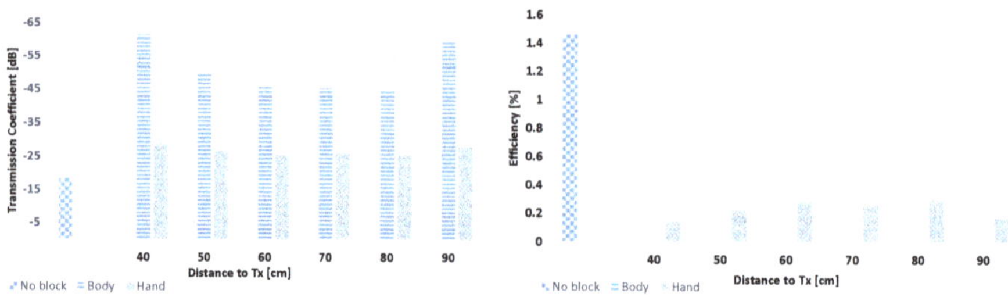

(a) (b)

Figure 15. Performance comparison between no block, block with a human body and hand (Rx is 100 cm away from the metasurface (Tx)): (**a**) transmission coefficient; (**b**) efficiency.

Table 2. Transmission coefficient enhancement as re-training with human body exposure.

Position	Transmission Coefficient (dB)		Improvement (dB)
	without Re-Training	Re-Training	
1. Close to Tx	−61.4	−54.86	6.54
2. In the middle	−46	−43.83	2.17
3. Close to Rx	−59.4	−54.5	4.9

Table 3. Comparison of the proposed system with previous works.

Reference	Operating Frequency	Dimensions	Method	Adaptability	Distance	Efficiency
[27]	920 MHz	$5.2\lambda \times 2.6\lambda$	Phased array	Yes	6 m	1.2%
[26]	5.8 GHz	$23.59\lambda \times 24.55\lambda$	Reflectarray	No	6 m	25%
Our work	5.8 GHz	$3.4\lambda \times 3.4\lambda$	Coding metasurface	Yes	1 m	1.5%

5. Conclusions

In this paper, a novel 1-bit coding metasurface is proposed and implemented for WPT systems. The coding metasurface is capable of adaptively focusing and steering beam to enhance power transfer efficiency in WPT systems. The proposed metasurface comprises 16 × 16 compact unit cells which were designed with a fractal structure and an operating frequency of 5.8 GHz. Indeed, both simulations and experiments were conducted to validate the theory. It was demonstrated that the proposed coding metasurface is able to focus the beam to the desired direction with a wide range from -60° to 60° in the elevation angle. For the adaptive beam to be focused on the mobile receiver, the adaptive optimal phase control scheme was proposed and applied. The results prove that the optimal one surpasses the random phase control and beam synthesis schemes. At 50 cm, the coding metasurface with the optimal phase control scheme can provide approximate 4% of power transfer efficiency, while it is just about 3% in the beam synthesis scheme.

Author Contributions: Conceptualization, N.M.T.; Methodology, N.M.T.; Software, N.M.T. and M.M.A.; Validation, N.M.T., M.M.A., J.H.P. and S.I.H.; Formal analysis, N.M.T., M.M.A.; Investigation, N.M.T., M.M.A. and J.H.P.; Resources, J.H.P., S.I.H.; Data curation, N.M.T.; Writing—original draft preparation, N.M.T.; Writing—review & editing, K.W.C. and D.I.K.; Visualization, N.M.T.; Supervision, K.W.C. and D.I.K.; Project administration, K.W.C. and D.I.K. All authors have contributed to the editting and proofreading of this paper.

Funding: This research was supported in part by the National Research Foundation of Korea (NRF) grant funded by the Korean government (MSIP) (2014R1A5A1011478), and in part by the MSIT(Ministry of Science and ICT), Korea, under the ITRC(Information Technology Research Center) support program(IITP-2019-2016-0-00311) supervised by the IITP(Institute for Information & communications Technology Planning & Evaluation).

Conflicts of Interest: The authors declare no conflict of interest.

Abbreviations

The following abbreviations are used in this manuscript:

WPT	Wireless Power Transfer
IoTs	Internet of Things
WSN	Wireless Sensor Network
EM	Electromagnetic
PAA	Phased Array Antennas
HPBW	Half Power Beamwidth
CST	Computer Simulation Technology
FCC	Federal Communications Commission

References

1. Tesla, N. Apparatus for Transmitting Electrical Energy. United States Patent US1119732A, 1 December 1914. Available online: https://patents.google.com/patent/US1119732A/en (accessed on 25 November 2019).
2. Xie, L.; Shi, Y.; Hou, T.; Lou, W. Wireless Power Transfer and Applications to Sensor Networks. *IEEE Wirel Commun.* **2013**, *20*, 140–145.
3. Kamalinejad, P.; Mahapatra, C.; Sheng, Z.; Mirabbasi, S.; Leung, V.C.M.; Guan, Y.L. Wireless Energy Harvesting for the Internet of Things. *IEEE Commun. Mag.* **2015**, *53*, 102–108. [CrossRef]
4. Liu, S.; Cui, T.J.; Xu, Q.; Bao, D.; Du, L.; Wan, X.; Tang, W.X.; Ouyang, C.; Zhou, X.Y.; Yuan, H.; et al. Anisotropic coding metamaterials and their powerful manupulation of differently polairzed terahertz waves. *Light Sci. Appl.* **2016**, *5*, e16076. [CrossRef] [PubMed]
5. Cui, T.J.; Qi, M.Q.; Zhao, J.; Cheng, Q. Coding metamaterials, digital metamaterials and programmable metamaterials. *Light Sci. Appl.* **2014**, *3*, e218. [CrossRef]
6. Campbell, S.D.; Sell, D.; Jenkins, R.P.; Whiting, E.B.; Fan, J.A.; Werner, D.H. Review of numerical techniques for meta-device desing [Invited]. *Opt. Mater. Express* **2019**, *9*, 1842–1863. [CrossRef]
7. Cui, T.J.; Liu, S.; Zhang, L. Information metamaterials and metasurfaces. *J. Mater. Chem. C* **2017**, *5*, 3644–3668. [CrossRef]

8. Yang, H.; Yang, F.; Xu, S.; Mao, Y.; Li, M.; Cao, X.; Gao, J. A 1-Bit 10 × 10 Reconfigurable Reflectarray Antenna: Design, Optimizatioin, and Experiment. *IEEE Trans. Antennas Propag.* **2016**, *64*, 2246–2254. [CrossRef]
9. Yang, H.; Yang, F.; Xu, S.; Li, M.; Cao, X. Experimental Study of a 1-Bit 10 × 10 Reconfigurable Reflectarray Antenna. In Proceedings of the IEEE International Symposium on Antennas and Propagation & USNC/URSI National Radio Science Meeting, Vancouver, BC, Canada, 19–24 July 2015; pp. 2153–2154.
10. Yang, H.; Yang, F.; Cao, X.; Xu, S.; Cao, J.; Chen, X.; Li, M.; Li, T. A 1600-Element Dual-Frequency Electronically Reconfigurable Reflectarray at X/Ku-Band. *IEEE Trans. Antennas Propag.* **2017**, *65*, 3024–3032. [CrossRef]
11. Yang, H.; Yang, F.; Xu, S.; Li, M.; Cao, X.; Gao, J. A 1-Bit Multipolarization Reflectarray Element for Reconfigurable Large-Aperture Antennas. *IEEE Antennas Wirel. Propag. Lett.* **2017**, *16*, 581–584. [CrossRef]
12. Yang, H.; Cao, X.; Yang, F.; Gao, J.; Xu, S.; Li, M.; Chen, X.; Zhao, Y.; Zheng, Y.; Li, S. A programmable metasurface with dynamic polarization, scattering and focusing control. *Sci. Rep.* **2016**. [CrossRef]
13. Tian, S.; Liu, H.; Li, L. Design of 1-Bit Digital Reconfigurable Reflective Metasurface for Beam-Scanning. *Appl. Sci.* **2017**, *7*, 882. [CrossRef]
14. Huang, C.; Sun, B.; Pan, W.; Cui, J.; Wu, X.; Luo, X. Dynamical beam manipulation based on 2-bit digitally-controlled coding metasurface. *Sci. Rep.* **2017**, *7*, 42302. [CrossRef]
15. Li, Y.B.; Li, L.L.; Xu, B.B.; Wu, W.; Wu, R.Y.; Wan, X.; Cheng, Q.; Cui, T.J. Transmission-Type 2-Bit Programmable Metasurface for Single-Sensor and SingleFrequency Microwave Imaging. *Sci. Rep.* **2016**, *6*, 23731. [CrossRef]
16. Ranaweera, A.L.A.K.; Pham, T.S.; Bui, H.N.; Ngo, V.; Lee, J.W. An active metasurface for feldlocalizing wireless power transfer using dynamically reconfgurable cavities. *Sci. Rep.* **2019**, *9* ,11735. [CrossRef]
17. Lang, H.D.; Saris, C.D. Optimization of Wireless Power Transfer Systems Enhanced by Passive Elements and Metasurfaces. *IEEE Trans. Antennas Propag.* **2017**, *65*, 5462–5474. [CrossRef]
18. Smith, D.R.; Gowda, V.R.; Yurduseven, O.; Larouche, S.; Lipworth, G.; Urzhumov, Y.; Reynolds, M.S. An analysis of beamed wireless power transfer in the Fresnel zone using a dynamic, metasurface aperture. *J. Appl. Phys.* **2017**, *121*, 014901. [CrossRef]
19. Song, M.; Baryshnikova, K.; Markvart, A.; Belov, P.; Nenasheva, E.; Simovski, C.; Kapitanova, P. Smart table based on a metasurface for wireless power transfer. *Phys. Rev. Appl.* **2019**, *11*, 054046. [CrossRef]
20. Zhang, P.; Li, L.; Zhang, X.; Liu, H.; Shi, Y. Design, measurement and analysis of near field focusing reflective metasurface for dual-polarization and multi-focus wireless power transfer. *IEEE Access* **2019**, *7*, 110387–110399. [CrossRef]
21. Yu, S.; Liu, H.; Li, L. Design of near-field focused metasurface for high-efficient wireless power transfer with multifocus characteristics. *IEEE Trans. Ind. Electron.* **2019**, *66*, 3993–4002. [CrossRef]
22. Li, L.; Liu, H.; Zhang, H.; Xue, W. Efficient wireless power transfer system integrating with metasurface for biological applications. *IEEE Trans. Ind. Electron.* **2018**, *65*, 3230–3239. [CrossRef]
23. Nadeem, Q.U.A.; Kammoun, A.; Chaaban, A.; Debbah, M.; Alouini, M. S. Intelligent Reflecting Surface Assisted Multi-User MISO Communication. *arXiv* **2019**, arXiv:1906.02360.
24. Yang, Y.; Zheng, B.; Zhang, S.; Zhang, R. Intelligent Reflecting Surface Meets OFDM: Protocol Design and Rate Maximization. *arXiv* **2019**, arXiv:1906.09956.
25. Mishra, D.; Johansson, H. Channel Estimation and Low-Complexity Beamforming Design for Passive Intelligent Surface Assisted MISO Wireless Energy Transfer. In Proceedings of the IEEE International Conference on Acoustics, Speech and Signal Processing (ICASSP), Brighton, UK, 12–17 May 2019; pp. 4659–4663.
26. Lipworth, G.S.; Hagerty, J.A.; Arnitz, D.; Urzhumov, Y.A.; Nash, D.R.; Russell, J. A Large Planar Holographic Reflectarray for Fresnel-Zone Microwave Wireless Power Transfer at 5.8 GHz. In Proceedings of the EEE/MTT-S International Microwave Symposium-IMS, Philadelphia, PA, USA, 10–15 June 2018; pp. 964–967. [CrossRef]
27. Aziz, A.A.; Ginting, L.; Setiawan, D.; Park, J.H.; Tran, N.M. Battery-Less Location Tracking for Internet of Things: Simultaneous Wireless Power Transfer and Positioning. *IEEE Internet Things* **2019**, *65*. [CrossRef]

energies

MDPI

Article

Energy-Efficient Topology Control for UAV Networks

Seongjoon Park †, Hyeong Tae Kim † and Hwangnam Kim *

School of Electrical Engineering, Korea University, Seoul 02841, Korea; psj900918@korea.ac.kr (S.P.); kevink97@korea.ac.kr (H.T.K.)

* Correspondence: hnkim@korea.ac.kr; Tel.: +82-02-3290-4821

† These authors contributed equally to this work.

Received: 24 October 2019; Accepted: 25 November 2019; Published: 27 November 2019

Abstract: Following striking developments in Unmanned Aerial Vehicle (UAV) technology, the use of UAVs has been researched in various industrial fields. Furthermore, a number of studies on operating multiple autonomous networking UAVs suggest a potential to use UAVs in large-scale environments. To achieve efficiency of performance in multi-UAV operations, it is essential to consider a variety of factors in UAV network conditions, such as energy efficiency, network overhead, and so on. In this paper, we propose a novel scheme that improves the energy efficiency and network throughputs by controlling the topology of the network. Our proposed network topology control scheme functions between the data link layer (L3) and the network layer (L2). Accordingly, it can be considered to be layer 2.5 in the network hierarchy model. In addition, our methodology includes swarm intelligence, meaning that whole topology control can be generated with less cost and effort, and without a centralized controller. Our experimental results confirm the notable performance of our proposed method compared to previous approaches.

Keywords: UAV network; topology control; space division; energy efficiency

1. Introduction

The biggest advantage of Unmanned Aerial Vehicles (UAVs) is three-dimensional mobility with a high degree of freedom, and the relatively low cost of the devices, which leads to the possibility of large-scale operation [1]. For instance, [2] expands the network infrastructure by using UAVs as an Access Point (AP) with mobility, and [3] deploys a scalable surveillance network with three-dimensional vision of UAVs. To operate these UAV applications well, there are numerous requirements regarding the networking. In particular, since the UAV network needs to be operated under various conditions, it needs to be resilient to dynamic changes of topology, intermittent links failures, resource constraints, three-dimensional mobility, equality on link replacement, and so on [4]. Furthermore, to fully use the multi-UAV large-scale, fast, and flexible mobile wireless network, designing a high-performance multi-hop UAV network is regarded as one of the core objectives of the UAV industry, which has been continuously addressed [5–8]. Compared to the importance of these network design trends, research into UAV networks has suffered from the lack of applicability or the reliability, due to the variety of environments and the hardware specifications. In particular, from the point of the view of energy efficiency of multi-hop UAV networks, there are serious leaks in the power consumption of communications, which also has the potential to degrade the throughput of the entire network.

In a conventional UAV network configuration, one UAV transmits the messages with the same power level to all UAVs in its transmission range. As shown in Figure 1a, a UAV makes the link connection to all UAVs which are in its transmission range, with the same transmission power P_{Tx}. Indeed, constructing *full* connections provides strong stability to the network. However, such topology with indiscriminate transmissions generates an inefficient UAV network by continuously generating more power consumption than actually needed, which highly reduces each UAV's network operating time. Also, the transmission power exceeding the minimal requirement on the links of the UAVs increases the possibility of the interference or unexpected silence of the wireless medium, which drops the channel use. Although there were some studies actively controlling the transmission power [9–11], the targeted network topology is constrained such as WLAN, and the existence of the centralized coordinators could limit the extensibility of multi-UAV operation. To resolve the power consumption problem, constructing a Minimum Spanning Tree (MST) of the network and minimizing the transmission power can be desirable. Figure 1b shows the graphical representation of the network topology, which is shaped like a MST. $P_{Tx,i}$ refers to the transmission power of i-th UAV, which is managed by the central or global controller. The root of the tree might be the gateway or the sink node of the UAV network. Although this centralized way can highly reduce power consumption with less routing overhead [12], its resulting topology can also bring connectivity issues. If all UAVs are connected by only a few paths, the energy consumption per hop is reduced but the overall network connectivity becomes unstable due to there being fewer options to route, and this is critical in UAV networks which have high mobility. Furthermore, increasing hop count can cause higher power consumption compared to a smaller hop count connection, and increase the forwarding overhead.

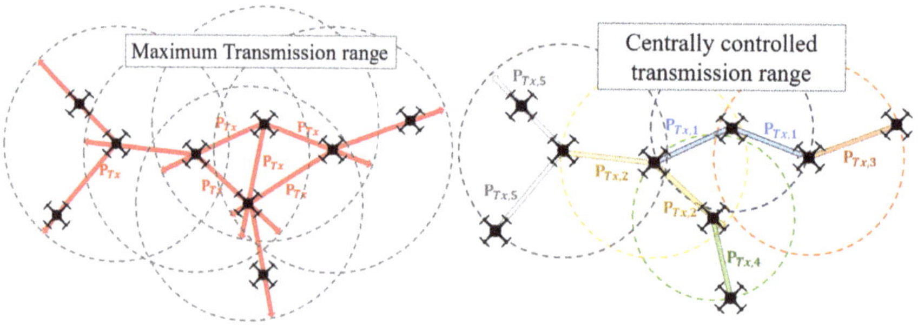

(a) (b)

Figure 1. Existing network topologies. (**a**) Conventional topology constructed by the T_x threshold; (**b**) Conventional topology constructed by the global MST.

To find the proper design of the network topology, we propose a novel distributed topology control scheme such that each UAV variably adjusts the transmission power while maintaining the efficient link connections through the space partition method. The core motivation of our concept is the intermediate layer design between the data link layer (L2) and the network layer (L3). As the network layer constructs a routing table using all its nearby UAVs, the number of the available links is too large in a dense environment, so some links might be inefficient due to the relatively large distance, interference, or hidden terminals. On the other hand, although the data link layer can control the transmission power, it cannot consider the packet-level power control by itself; the data link layer does not know the proper power for each hop. If a layer that designs the *topology* of the network prunes the links and determines proper transmission power, then it can help the network layer design a more efficient routing table, while making the data link layer transmit with more efficient power. We called this concept of the intermediate layer the *topology control layer (L2.5)*. By explicitly controlling the available links of the UAVs, it can reduce the power consumption while maintaining the robustness of the network connection.

Pruning the links in the topology control layer is based on the space partition, to gain the advantage in the link cost. Figure 2 shows the graphical representation of our proposed topology control scheme. Periodically, each UAV equally divides its transmission space into several partitions. Then, from each partition, the UAV picks one or more nearby UAVs and removes the other links. In Figure 2, $P_{Tx,k}$ refers to the determined transmission power of the link between the UAV itself and the neighbor UAV, which is one of the nearest ones in the k-th quadrant. By doing so, the UAV network can maintain its topology where each UAV has multiple links with nearby neighbors, which has a chance to dynamically control the transmission power at each hop.

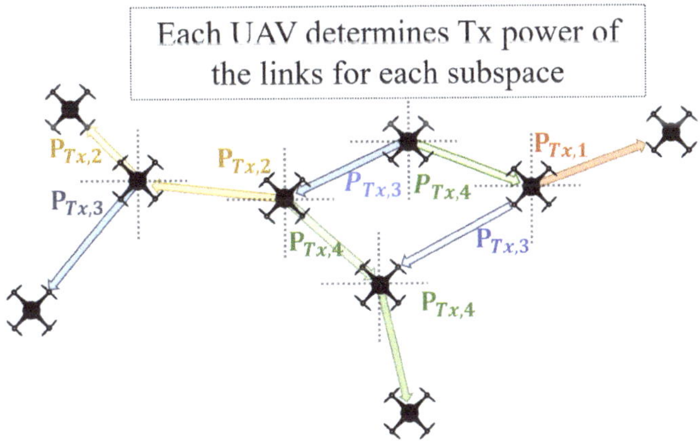

Figure 2. Example of topology control layer with 4 partitions.

Based on the suggested topology control method, we implemented the simulation to legitimize the performance improvements in the UAV network. From the numerical analysis, it is shown that topology control layer has a gain in throughput, network stability, and energy efficiency in the UAV network. Our contribution can be summarized as follows.

- With our proposed network topology control method, a UAV network can be formed with energy-efficient network properties.
- Our scheme does not require the UAV or the Network Manager to know the whole information about all UAVs, such as position, routing path, and so on. Instead, by using the concept of *swarm intelligence*, each UAV is only concerned about their next one-hop connection in each partition, which will consequently make the effective network topology.
- The number of the partitions can be varied for each UAV. The number of the divided spaces can be customized with network density.
- Our research can be compatible with other network layers in the OSI model. Also, the topology control layer acts just between the data link layer (L2) and network layer (L3), which possibly can be expressed as *L2.5*.

In the following sections, we provide further explanation of our research. In Section 2, we introduce previous research on UAV networking technology. In Section 3, we present our methodology that explains the system overview. The evaluation of our methodology is presented in Section 4, and Section 5 lists the discussions about the simulation results. Finally, we conclude our work in Section 6.

2. Related Work

Since UAV networking is an intensive technology that requires a variegated research domain, various research is being actively conducted in several ways. In this paper, recent research focusing on the various topics on UAV formation, energy-efficiency design, and the transmission power control is introduced in detail.

- **UAV formation**: The construction of an appropriate formation with many UAVs is progressing in various aspects. In fact, UAV positioning algorithms have been studied in consideration of network throughput [13], obstacle collision avoidance [14], and mission conditions [15]. Sabino et al. [16] proposed the multi-UAV placement scheme through the genetic algorithm. It should be noted that UAV formation control algorithms can be joined with our proposed system, since our system aims to improve the network topology of UAVs, which have the potential to frequently change their shape in three-dimensional space. With our dynamic topology control scheme, the UAV network can sustain an energy-efficient network while changing each UAV's position according to their mission.

- **Energy Efficiency**: Research to improve the energy efficiency of the UAV network is proceeding by way of reducing the energy in the operation of UAV itself (e.g., optimized path planning for efficient aircraft propulsion), and the arrangement to minimize the energy on the entire network (e.g., optimal network recovery with additional UAV) [17]. In this paper, we compose an energy-efficient topology that is based on network management, which can be applied a priori to other related previous research.
- **Transmission Power Control**: Many studies have been made to change transmission strength according to its conditions for effective communication. For example, power-controlled multiple access MAC protocols (PCMA) were presented to improve the channel use [18], and a tunable circuit system was studied to generate range-adaptive transmit signals [19]. However, these studies have not been used for UAV networks with mobility characteristics. Moreover, these studies are still inefficient since all nodes in the transmission range attempt to connect without any adjustment. However, our research uses the transmission power control in the UAV network and forms the network by controlling the number of the available links.
- **3D topology control schemes**: Emerging research on the unmanned vehicle system emphasizes the importance of the dynamic network control problem for expanding its usage. Continuously adapting the network topology in the 3D space has been largely studied for both energy efficiency and network quality. Zhang et al. [20] proposed a cluster sleep–wake scheduling algorithm for underwater sensor networks. One of the main differences with our system is that this algorithm produces an on–off schedule of the links, while our one results in selected links which guarantee global connectivity (Section 3.3). Also, it adopts a centralized partitioning concept, while each UAV partitions the space, centering itself, in our algorithm. Our distributed concept has the advantage of scalability of the network, since the processing overhead remains the same with larger-scale networks. Kim et al. [21] addresses the 3D topology control method considering interference. In this paper, Cone-Based Topology Control (CBTC) has been introduced for partitioning each node's space. The major difference with our work is the group of the selecting nodes. Our system selects the adjacent UAVs of the MST constructed in each partition (Section 3.2), while CBTC selects only the nearest one in each partition. This difference shows the limitation of the number of partitions, where CBTC restricts the angle of the division section while our system does not. The flexibility of the number of partitions results in the variety of the partition models, which enlarges the adaptability of the network scale (e.g., density of the UAVs).

In summary, there has been research considering the energy consumption of the wireless UAV network. However, due to the lack of the cooperability and the compatibility with the other OSI layers, the UAV network is not fully beneficial when previous strategies are applied. In addition, some attempts at controlling the mobility of the UAV for networking energy efficiency might not be helpful when operating the UAVs in a practical scenario, since each UAV's position decision will suffer from the confliction between the power efficiency and the mission performance. Our design adopts the distributed and reactive method, where each UAV collects neighbors' locations and determines their available links and the transmission power for energy-efficient networking.

3. System Design

In this section, we address the design of the topology control layer. Our concept is an additional layer between the data link layer and the network layer, so it coordinates these layers to improve the energy efficiency of the network, by controlling the topology. Also, our design targets the distributed manner of the system, whose result guarantees a topology with high efficiency and complete connectivity. The following subsections explains our system components in detail.

3.1. Topology Control Layer

The topology control method that we propose is individually applied to each UAV in the network separately, but the final network appears as a comprehensive network topology. As shown in Figure 3, our topology control process is functioning between the network layer (L3) and the data link layer (L2). Since the topology control is compatible with various routing protocols in the network layer and many other flow control models in the data link layer, it can be well used in the conventional OSI network system. We assume that each UAV is equipped with a global navigation module, such as GPS, to determine the relative position between the UAVs. The topology control layer periodically broadcasts its position information, so each UAV can seize the nearby UAV positions. For example, as shown in Figure 3, let us assume that a UAV collects the positions of the four neighbor UAVs, from the beacon message reception by L2. The topology control layer *filters out* the link among these four available links, such as L_2 and L_3, and reports the information of the link L_1 and L_4 to the L3. L3 operates its routing algorithm and creates the routing table. After L3 determines the link to send the packet, the topology control layer forwards the designated transmission power to L2, so the packet reaches the next hop or the destination with a proper amount of signal power. Note that if $P_1 < P_4$, then there is a possibility that UAV 0 does not interfere with UAV 1 while using L_4, which has advantages in the congestion release of the wireless medium.

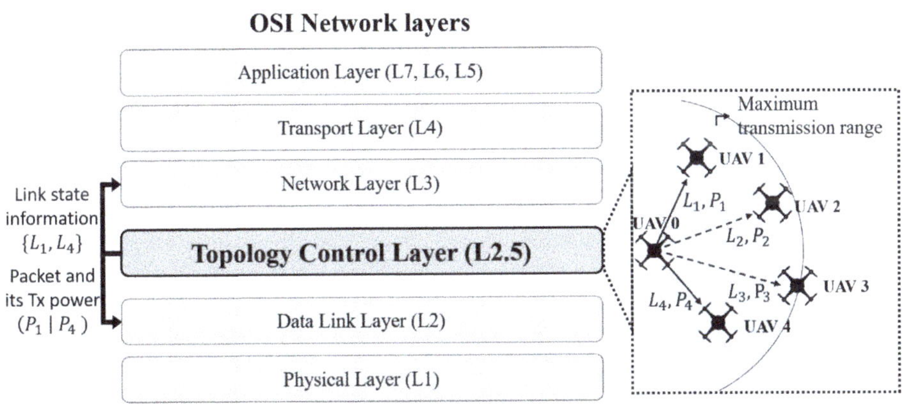

Figure 3. Topology control layer as *L2.5* in the Network OSI model.

Compared to the fully connected topology (e.g., Figure 1a), some advantages can be made when adopting topology control layer. First, transmission power per each link is adapted. Of course, enough of the transmission power is needed for higher throughput, but excessively large power can be redundant at the viewpoint of the network lifetime. Second, a routing table is simplified and more effective. Due to the link filtering of the topology control layer, the number of the routes decreases, so the routing table calculation overhead also decreases. On the other hand, compared to the global MST topology (e.g., Figure 1b), there are other improvements, as in the case of fully connected topology. Due to the packet-level power control, each UAV can handle more links than the MST case. Also, there are more capabilities to cope with the congestion or the failure of the links, since the topology control layer periodically observes the neighboring UAVs and considers the substitute route for networking.

Now, we should consider how to filter the links from the available ones. If the criteria of the filter are only the distances from the destination UAVs, then a part of the UAV network can be isolated due to unexpected link pruning, as shown in Figure 4. Section 3.2 discusses our strategy to build stabilized and efficient topologies by collecting only the positions of nearby UAVs.

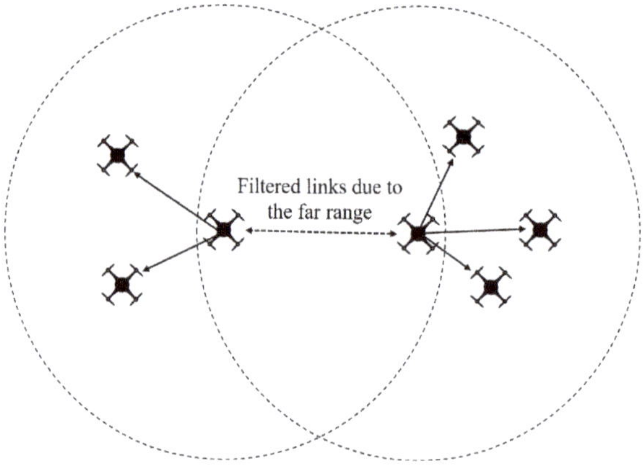

Figure 4. Wrong case of the link pruning.

3.2. Space Partition Method

To distribute the neighbor links direction evenly, we propose a novel space partition method as a solution of the link selection problem. Assuming that the maximum transmission range of a UAV is shaped as a sphere, the spherical space can be divided into several partitions with the same volume, which includes the origin of the sphere. By selecting the closest links from each partition, the UAV gets a set of links that are evenly distributed in all directions. With this observation, we designed an algorithm for the proposed space partition scheme as shown in Algorithm 1.

Figure 5 shows the overall steps of Algorithm 1. Our proposed system runs two processes in parallel, the one addressed at lines 3 to 11, and the other one addressed at lines 13 to 32. The system globally allocates a memory to monitor the location of the nearby UAVs in the maximum transmission range, named P. The function `NeighborUpdation` periodically broadcasts the beacon message containing its position information, and receives the nearby UAV's beacon message to update P. With the updated position information, the function `LinkSelection` prunes the links according to the space partition method and designates the transmission power of each link. We address the sequence of the algorithm in detail.

At first, the algorithm should secure a set of vectors S, which contains the vectors, each of which projects the partitioned surface of a sphere, known as *partition vector*. To get S, we divide the sphere sized with the maximum transmission range into n partitions. For the uniform distribution of the links, we set ground rules for partitioning:

- The partitions should have the same volume.
- The partitions should contain the origin of the sphere.
- The partitions should have the same area of the surface.

Algorithm 1 Topology control layer.

$P \leftarrow \phi$ ▷ Neighbors' position vectors

function NEIGHBORUPDATION
  while exception occurred **do**
    $\overrightarrow{p_0} \leftarrow$ `UpdatePos()`
    `BroadcastPosition(`$\overrightarrow{p_0}$`)`
    $P' \leftarrow$ `ReceiveBeaconMessages()`
    Update P with P'
    Sleep for a cycle
  end while
end function

function LINKSELECTION
  $S \leftarrow$ Normal vectors of n planes. ▷ Partition vectors
  while exception occurred **do**
    $A_i \leftarrow \phi$ where $0 < i < n$
    $H_i \leftarrow \phi$ where $0 < i < n$
    $Tx_i \leftarrow 0$ where $0 < i < n$
    for $\overrightarrow{p}$ **in** P **do**
      $\overrightarrow{d} \leftarrow \overrightarrow{p} - \overrightarrow{p_0}$
      $\overrightarrow{s}^* \leftarrow \arg\max_{\overrightarrow{s} \in S} \overrightarrow{s} \cdot \overrightarrow{d}$
      $h \leftarrow$ `IndexOf(`$\overrightarrow{s}^*$`)`
      append `IndexOf(`$\overrightarrow{p}$`)` **in** A_h
    end for
    for i where $0 < i < n$ **do**
      Construct an MST of the components in A_i regarding $\overrightarrow{p_0}$ as a root
      $H_i \leftarrow$ the children of $\overrightarrow{p_0}$
      $Tx_i \leftarrow$ `GetTxOfRange(`H_i`)`
    end for
    Sleep for a cycle
  end while
end function

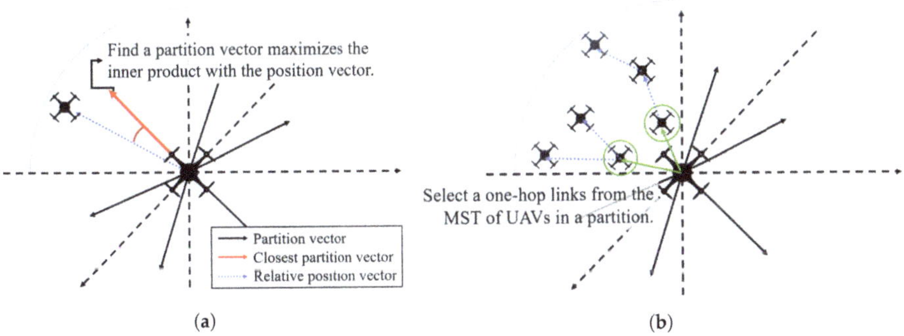

Figure 5. Graphical representation of Algorithm 1. (**a**) Finding the belonging partition of a link; (**b**) Select the nearest link from each partition.

The examples of the partitioning are represented in Figure 6, where $n = \{2, 4, 8, 12, 20\}$. For visibility, we expressed the partitions as the shape of the original ingredients of the solid figure, such as cube, icosahedron, and so on. The actual partitions are the ones trimmed by the sphere inscribed in the colored solid. For instance, if $n = 2$, a set S is composed of two vectors, which are $(1,0,0)$ and $(-1,0,0)$, respectively. On the other hand, if $n = 8$, a set S is composed of eight vectors, which are $(\pm 1, \pm 1, \pm 1)$. In the cases of $n = 12$ and $n = 20$, we used known regular polyhedrons which have the same area of the faces and can be closely fitted to the sphere, such as dodecahedron and icosahedron, respectively. The normal vectors of each case can be derived from the known equations, such as

$$\begin{gathered} S = \{(\pm 1, \pm 1, \pm 1), \\ (0, \pm(1+h), \pm(1-h^2)), \\ (\pm(1+h), \pm(1-h^2), 0), \\ (\pm(1-h^2), 0, \pm(1+h))\} \\ \textit{where } h = \frac{\sqrt{5}-1}{2}, \end{gathered}$$

when $n = 20$.

Topology control layer periodically updates the position of nearby UAVs by `NeighborUpdation`, and determine what partition each UAV belongs to, in `LinkSelection`. For each UAV's position $\overrightarrow{p}$, the algorithm derives $\overrightarrow{d} = \overrightarrow{p} - \overrightarrow{p_0}$, and selects a normal vector s^* from S which results the maximum inner product value with $\overrightarrow{d}$ (Figure 5a). After categorizing all links, the algorithm constructs an MST composing the UAV itself and the other UAVs in each partition, then collects the one-hop links from them (Figure 5b). Finally, the UAV reserves the number of the selected links, each of which is the nearest UAV in a partition. In the aforementioned case of Figure 4, the far-range link will be one of the selected one, since the neighbor UAV is within a transmission range, and it will be the only one link of a specific partition.

After the suggested topology control process, multi-hop communication can be performed with any desired routing protocol. For instance, the shortest path algorithm, such as the Dijkstra or Bellman–Ford algorithm, can be used to derive the routing path to the other UAV connections in distributed routing protocols. As an overall assessment, our topology control method can reduce the number of unnecessary links by effectively dividing the surrounding space into several partitions. Also, it has an advantage in transmission power because each UAV only needs to consider the nearest UAVs.

3.3. Connectivity Proof

We show the connectivity of the resulting topology of Algorithm 1 derived from the given deployment of UAVs. We preliminarily assume that any of the deployed UAVs has one or more nearby UAVs in its maximum transmission range. We first prove local connectivity among the neighbors in the maximum transmission range of UAV. Then, we eventually prove global connectivity by finding a knock-on path from two arbitrary UAVs u to v, referred to $P(u,v) = < u, h_1, h_2, h_3, \ldots, h_m, v >$. This global connectivity is significantly derived from UAVs' previous local connections. In terms of the resulting topology of our system, we derive the following two theorems about connectivity.

Theorem 1 (Local connectivity). *The result of Algorithm 1 guarantees the connectivity between a UAV and its neighbor UAVs within the maximum transmission range.*

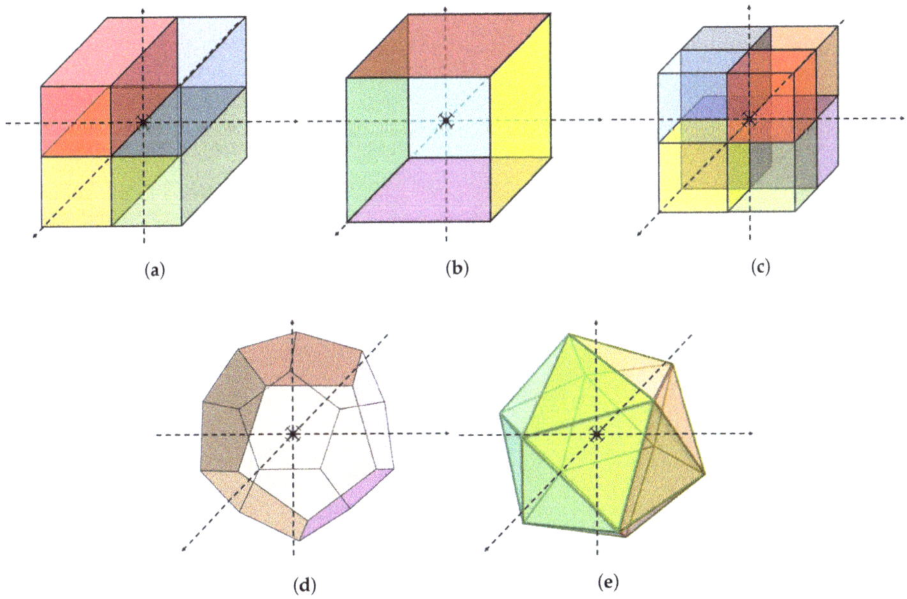

Figure 6. Space partition with respect to n. (**a**) Space partition when $n = 4$; (**b**) Space partition when $n = 6$; (**c**) Space partition when $n = 8$; (**d**) Space partition when $n = 12$; (**e**) Space partition when $n = 20$.

Proof of Theorem 1. This theorem shows the UAV connectivity between an arbitrary UAV u and its neighbor UAVs, all of which are within one maximum transmission range from the UAV u itself. Let MTR_u refer to a set of the UAVs located within the maximum transmission range, where $w \in MTR_u$. Also, let NH_u refer to a set of selected links by u in Algorithm 1, where $NH_u \subset MTR_u$.

$\forall w \in MTR_u$,

Case (i) : $w \in NH_u$

$P(u,w) = <u,w>$

Case (ii) : $w \notin NH_u$

there always exists w_1 such that $w_1 \in NH_u$ and

$P' = <u,w_1>$ is a subpath of $P_u(u,w)$,

which is derived by an MST constructed in line 26 of Algorithm 1.

We let $u = w_0$, and applying Algorithm 1 to each w_i,

then there exists w_{k-1} such that $w_k = w$.

By chaining all the discovered subpaths,

$P_{Total} = P'(w_0(=u), w_1) \cup P''(w_1, w_2) \cup ... \cup P^k(w_{k-1}, w_k(=w)) = P(u,w)$.

Thus, the path $P(u,w) = <u, w_1, w_2, \ldots, w_{k-1}, w>$ exists.

□

Figure 7 graphically represents the above sequence. By this procedure, the connectivity with all UAVs in every partition of arbitrary chosen UAV u is always guaranteed. Thus, the connection is guaranteed for all UAVs in the maximum transmission range of UAV u.

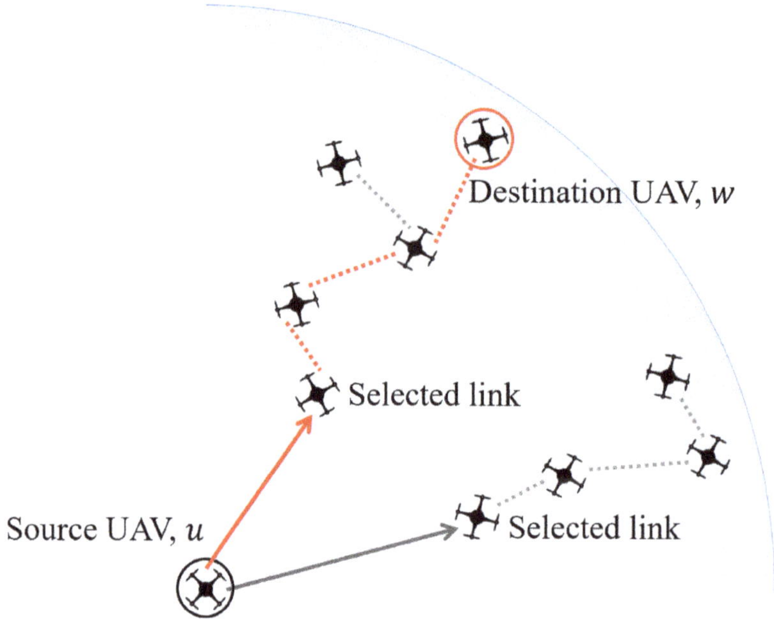

Figure 7. Finding routing path of the topology control layer.

Theorem 2 (Global connectivity). *The result of Algorithm 1 guarantees the connectivity of two arbitrary UAVs network-wide.*

Proof of Theorem 2. We prove the global connectivity with Theorem 1. We let $G(V,E)$ refer to the graph representing a fully connected UAV network, where V is the set of UAVs and E is the set of edges defined by the reachability between two UAVs with the maximum transmission range. Also, $G_{TC}(V,E_{TC})$ refers to the graph representing the resulting network topology of topology control layer, where E_{TC} is the set of edges that Algorithm 1 selects. The other notations mean the same as aforementioned.

Assume there is a path $P(u,v) = < u(= h_0), h_1, h_2, \ldots, v(= h_n) >$ in $G(V,E)$.
Each pair of subsequent UAVs in $P(u,v)$, such as $(h_i, h_{i+1})(i = 0, 1, \ldots, n-1)$,
is replaced with a path $< h_i, w_1, w_2, \ldots, w_{k-1}, h_{i+1} >$,
owing to Theorem 1.
Therefore, if $P(u,v)$ is in $G(V,E)$, then $P(u,v)$ is also in $G_{TC}(V,E_{TC})$.

□

3.4. Swarm Intelligence Point of View

We designed our proposed topology control to inherently contain the concept of swarm intelligence (SI). We made a simple rule for each UAV in the network, but the overall topology consequently shows robust connectivity, as proven in Section 3.3. Also, the ingredients to make decisions can be fairly collected in a distributed manner, where every UAV equally broadcasts its location. With changes to nearby environments, the UAV can dynamically change its link selections and designate the proper transmission power of the targeted end-to-end connections. Our accomplished SI concept yields the following advantages.

- **Scalability.** Compared to centralized methods, our proposed system does not need any centralized protocols or regulations. This feature prominently appears when the maximum hop count increases due to a large-scale network, where farthest UAVs take much longer to update its network configuration in the centralized system. In the topology control layer, each UAV determines the next available hops by itself, so the network size does not affect the network configuration delay.
- **Mobility.** Considering the high mobility of the UAV, our distributed concept based on the location has strong advantage during network topology changes. Since each UAV periodically updates its available neighbors, a network layer can rapidly drop or append the available links without hesitation.
- **Simplicity.** In the case of the micro UAVs, the computational resource is too small to input the high intelligence necessary for networking [22]. Our proposed system is composed of simple calculations augmented at the existing network stack, so it has much less occupancy of the system resources, which also contributes to the energy efficiency.

4. Evaluation

We implemented our proposed space partition method (Algorithm 1) and the network topology simulation written in Python, and measured the numerical data to plot the results using MATLAB. We used OpenCV library to display the network topology for various cases. Also, we compared our scheme with the following network topologies to present the advantages of applying the topology control process to the UAV fleet network.

1. **Fully connected (FC)**: UAV networking that is fully connected with all UAVs that are in their transmission range.
2. **Simple MST (SMST)**: UAV networking appending the simple minimum spanning tree method.
3. **Topology Control (TC)**: UAV networking with a proposed topology control method. For each n value, we abbreviate the TC with n partitions as *TC-n*.

To examine the performance of the resulting topologies, we simulated a routing scenario for each case. We randomly sampled 50% of the existing UAVs, and searched the optimal route from each to all the other ones. We adopted Dijkstra algorithm [23] to find the shortest path to the target destinations. As mentioned in Section 3.1, our topology control algorithm can be any other path-finding algorithms, such as congestion-free ad hoc routing strategies discussed in [24]. Also, we set the distance between the UAVs as a link cost used in our Dijkstra algorithm, so the result of the algorithm is the most energy-efficient paths with respect to the topology of FC, SMST, and TC-n.

We claim that our exhaustive search can thoroughly validate the performance of the network topology, along with the stability, energy efficiency, and the network traffic. Commonly used routing protocols aim to optimize the routing table of each device. Analysis of the Dijkstra algorithm result shows the best path of each end-to-end connection derived from the network topology, so we derive the statistics from the optimal path to every UAV in the network. We evaluate the network topologies by the following metrics.

- **Node degree:** We exploit one of the general terms used in the graph theory. In this paper, the term *node degree* refers to the number of the edges which are incident to a UAV. At the viewpoint of the network topology, high node degree implies high *stability* of the network. If one of the connected neighbor UAVs fails (due to the emergency landing or return to home), the UAV should use the other connections to sustain the network connectivity. However, if UAVs have low degrees, the network has higher potential to lose the whole connectivity even with the loss of some centric nodes.
- **Hop count:** Hop count of an end-to-end connection is the number of the edges of the optimal path between them, which can be derived by the aforementioned Dijkstra algorithm. Higher

hop count not only increases the delay of the connection but also has the potential to drop the end-to-end throughput, since the packet is repeatedly propagated through the wireless medium per each hop. Thus, lower hop count results in less use of the wireless medium with low latency, which results in the overall throughput improvement of the UAV network.

- **Power consumption:** As discussed in Section 3, our proposed system determines the transmission power of each link. We summed the amounts of the transmission power required at all links on each end-to-end connection. In the case of FC, we assumed there is no power control method equipped, so the expected power consumption is the multiplication of the average hop count by the maximum transmission power. On the other hand, in other cases, the transmission power of each link is calculated from the distance using the Friis equation. Please note that excessively high hop count results in the higher power consumption of the end-to-end communication, despite the low power consumption due to the short distance of the links.

We evaluate the network topologies while varying the number of UAVs, maximum transmission power, shape of the formation, and the number of the partition n. Default value of each parameter is listed in Table 1. The following subsections discuss the evaluation result while varying the parameters.

Table 1. Simulation parameters.

Item	Value
Space size	1000 m × 1000 m × 1000 m
Number of UAVs	50
Maximum transmission power	20 dBm
Frequency band	2.4 GHz
Antenna gain	2.5 dBi
Receive signal threshold	−70 dBm

4.1. Regular Formations

To effectively show the resulting shape of the network topology of our system, we first conducted the evaluation with the regular formation of the UAVs, such as grid-shape or sphere-shape. Figure 8 shows the three-dimensional representation of each network topology. In the case of the grid formation, we deployed $4 \times 4 \times 4$ UAVs with the default size of the map space, and the same distance of the width, height and depth between the nearby UAVs. In the case of the sphere formation, we deployed 66 UAVs in the surfaces of 3 concentric spheres. As shown in the Figure 8a, if all UAVs fully connect to the nearby UAVs in its maximum transmission range, the topology gets highly complex and this may lead to high interference in the wireless medium. On the other hand, the case shown in Figure 8b shows an MST including the network, which could lead to the high hop count of some connections, such as the route from 51 to 46, which has 17 hops. The topology result of our proposed topology control layer is shown in Figure 8c, which represents visually expected connections at this grid formation. The reason for the result is that the topology control layer only leaves the nearest link from each partition. While the fully connected cases in Figure 8a,d suggest the intensive traffic on the center of the network, the TP-6 cases in Figure 8c,f reduce this potential by filtering the further connections of each UAV and controlling the transmission power of each link. As shown in the topologies in Figure 8, we showed how our proposed system forms the topology of the UAV network, compared to the other comprehensive or centralized methods.

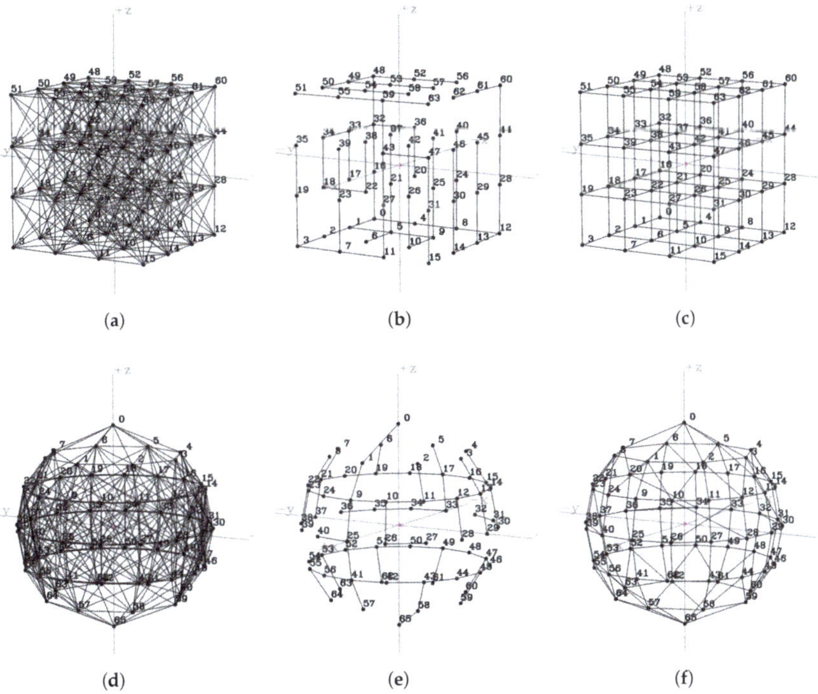

Figure 8. Network topologies in regular formations. (**a**) FC network topology of grid formation; (**b**) SMST network topology of grid formation; (**c**) TC-6 network topology of grid formation; (**d**) FC network topology of concentric spheres formation; (**e**) SMST network topology of concentric spheres formation; (**f**) TC-6 network topology of concentric spheres formation.

Figure 9 shows the evaluation results of the network topologies shown in Figure 8. We measured the average and the confidence interval of the power consumption, hop count of end-to-end connections, and the node degree of the UAVs. As shown in Figure 9a, TC-6 outperforms the other cases FC and SMST, while showing less than a half of the power consumption of the other cases. It is remarkable that TC-6 shows much less power consumption than SMST, which grants centralized optimal topology of the whole UAV network. The reason for this outperformance is mainly due to the advantage in the hop counts (Figure 9b), since larger hop count of the SMST case leads to more frequent transmissions, which incurs the large amount of the power consumption despite the low transmission power. By numerical comparison, we showed our proposed topology control layer can reduce the power consumption of the UAV network, through the efficiently constructed network topology. In the following graphs, for better visibility, we omitted the confidence interval of the results, which shows almost similar tendencies to the remaining evaluations.

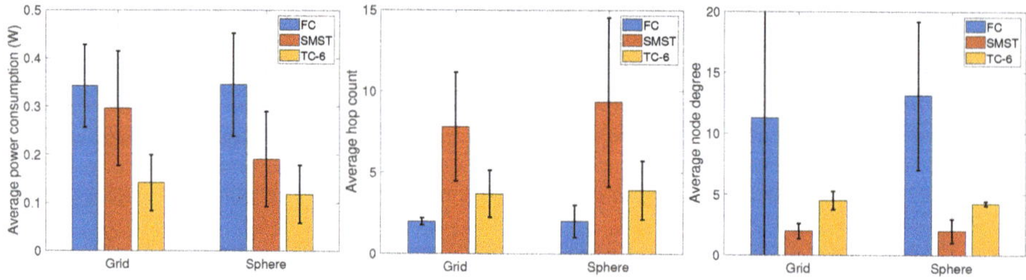

(a) (b) (c)

Figure 9. Numerical results in regular formations. (**a**) Average power consumption; (**b**) Average hop count; (**c**) Average node degree.

To show the differences in the network topology while varying n at the general case, we ran our proposed system in a random formation, as shown in Figure 10. In this figure, we deployed 20 UAVs in the map with uniform random distribution. As shown in the subfigures, when n increases, the shape of the network topology gets closer to the FC case, as TC-20, shown in Figure 10f. Please note that UAVs select at most n UAVs as their next hops in each partition, which has smaller size at larger n. The case of larger n has more chance to grab the UAVs in the transmission range as their next hops, so larger n acts as similar to the FC case. However, the expected power consumption of the end-to-end connection is smaller than the FC case even though larger n cases, since topology control layer designates the transmission power of the links. In the following subsections, we evaluate our system while varying the numerical parameters, to show how the parameters affect the performance of the topology control layer.

4.2. Network Size

Figure 11 shows the average of the power consumption, hop count and degree while varying the number of the UAVs from 20 to 100. We compared the case of FC, SMST, and TC-4 for each performance metric. The other parameters for the simulations are set as default, listed in Table 1. As shown in Figure 11a, TC-4 shows higher decrement ratio of the power consumption than the other cases. This is due to the unique property of the topology control layer, which adopts the transmission power of each link, while sustaining the number of the nearby links. Increasing the number of the UAVs leads to the decreases of the average distances between the UAVs. In the case of SMST, this decrement results in the low transmission power of the links just as TC-4, due to centralized topology generation. However, due to the properties of the minimum spanning tree, higher density of UAVs inevitably causes higher hop counts, which is shown in Figure 11b. As mentioned before, increasing hop count consequently increases the power consumption of the end-to-end connection, which results in relative inefficiency as in Figure 11a. On the other hand, the case of FC sustains smaller hop count and larger degree (Figure 11c) than the other cases, but it has much higher power consumption due to the fixed and excessive amount of the transmission power. Furthermore, a larger degree of UAVs indicates higher robustness, but also brings the potential for more congestion and the collision of the frames. Figure 11c shows that the case of FC steadily increases the degree of UAVs, which leads to the degradation of the throughput and the efficiency of the transmission. In summary, the results in Figure 11 proved that the topology control layer yields a moderated degree and hop count, which results in it outperforming the power efficiency of the UAV network.

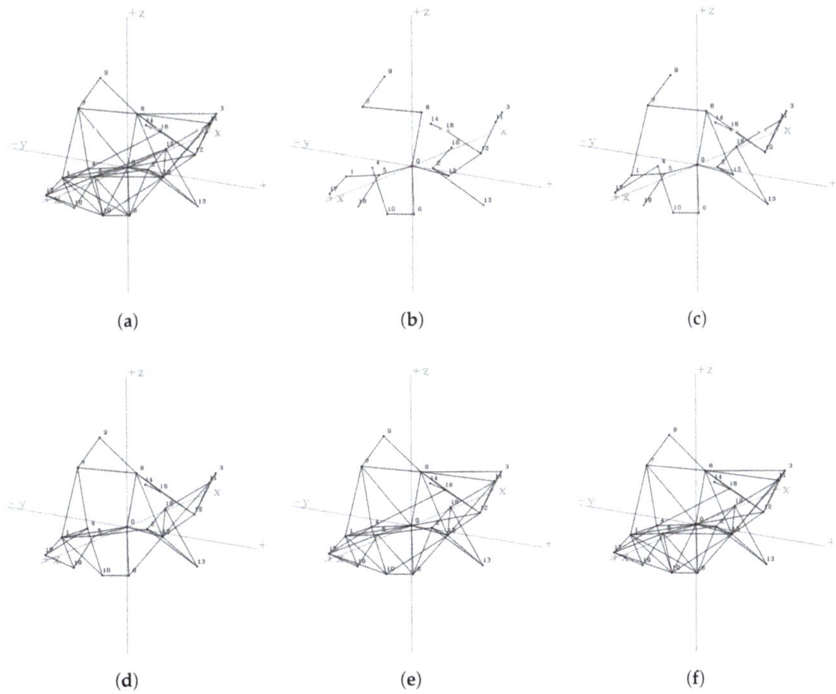

Figure 10. Network topology of a random formation while varying n. (**a**) FC network topology of random formation; (**b**) SMST network topology of random formation; (**c**) TC-2 network topology of random formation; (**d**) TC-4 network topology of random formation; (**e**) TC-8 network topology of random formation; (**f**) TC-20 network topology of random formation.

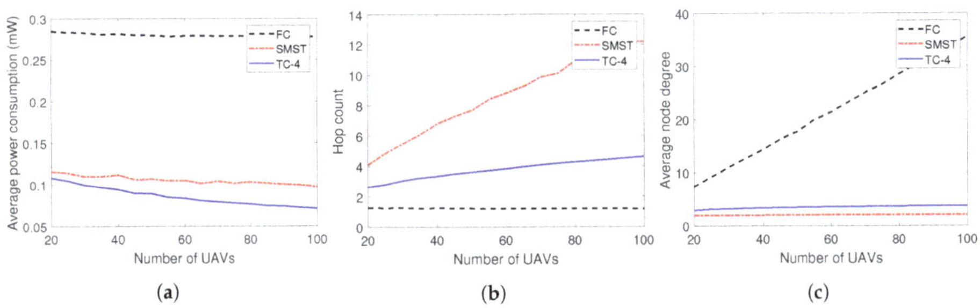

Figure 11. Numerical results while varying the number of UAVs. (**a**) Average power consumption; (**b**) Average hop count; (**c**) Average node degree.

4.3. Number of Partitions

We also measured the results while varying the number of the partitions, where $n = \{2, 4, 6, 8, 12, 20\}$. As shown in the subfigures in Figure 12, the cases of TC-n outperform the other cases (FC, SMST) in power consumption. In the case of the hop count and the degree, TC-n shows the medium values between the FC and the SMST, since the topology control layer prunes the links regarding to the number of the partitions. On the other hand, as n increases, hop count decreases and the degree increases. Due to the increase of the possible links in larger n, there are more chances to decrease the hop count with the larger number of partitions. Furthermore, it is notable that the average power consumption of end-to-end connection is minimum at $n = \{6, 8\}$, regardless of the number of the UAVs. However, it is hard to conclude that the network is *optimal* at a certain n, since the larger n results in lower hop count, which leads to the improvement of the expected throughput. In conclusion, the results in Figure 12 show that there is an optimal n value for the desired objective of the UAV network.

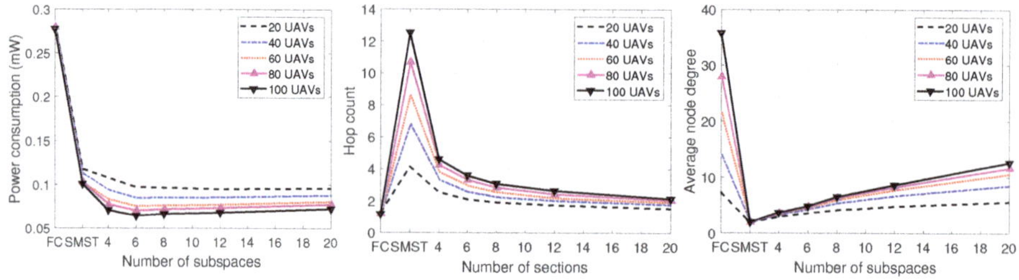

(a) (b) (c)

Figure 12. Numerical results while varying n. (**a**) Average power consumption; (**b**) Average hop count; (**c**) Average node degree.

4.4. Transmission Range

We varied maximum transmission power of 50 UAVs, in various n values. Figure 13 shows the numerical results compared to FC and SMST cases. In Figure 13a, we cut the transmission power over 0.5 mW for visuality. The results in this figure have several considerations. At first, while FC cases have exponentially increasing power consumption when the maximum transmission power increases, the cases of SMST and TC-n have mostly constant power consumptions. This result indicates that SMST and TC-n can sustain similar efficiency in various cases of the maximum transmission power. Then, the outperformance of the TC-n against the SMST originates from the packet-level power control and the partitioning-based link pruning, as mentioned in the former subsections. Secondly, there is a saturation while increasing the maximum transmission power, after about 22 dBm, both in the hop counts and the degrees. The reason for this phenomenon is that the nearest neighbors of certain directions are selected despite the transmission range expansion. For this reason, the saturation starts earlier when n is smaller, as shown in Figure 13b,c. The saturation indicates that our topology control layer outperforms results even at the smaller transmission range.

We summarize our simulation results as follows.

- We evaluated our proposed system *topology control layer* while varying the network size, number of partitions, and the transmission power. In every evaluation, our system shows outperforming results in average power consumption, while sustaining moderate values of the average node degrees and the hop counts.
- While varying the network size, the topology control layer shows a larger decrement in power consumption, because our system keeps a proper number of links and fits the transmission power of them.
- While increasing the number of the partitions, hop count shortly decreases and degree shortly increases due to the larger number of the selected links. In Figure 12a, TC-6 and TC-8 nearly show the lowest amount of the average power consumption regardless of the number of UAVs.
- Increasing the transmission range results in great advantages in power consumption, compared to the FC case. Also, we found that power consumption of lower n is relatively small in a large transmission range, while the consumption of higher n is relatively small in small transmission ranges.

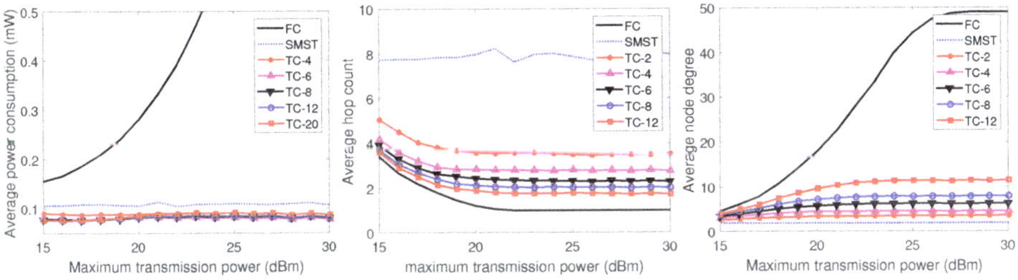

(a) (b) (c)

Figure 13. Numerical results while varying maximum transmission power. (**a**) Average power consumption; (**b**) Average hop count; (**c**) Average node degree.

From the aforementioned statements, we claim that the topology control layer can contribute to the energy-efficient communication in the UAV network, in wide range of the environments.

5. Discussions

This section addresses additional discussions for topology control layer. We listed our considerations as following.

- **Multi-input, Multi-output (MIMO) adaptation.** MIMO targets concurrent communications in several directions, which fits our concept well. Through the combination with MIMO technology, each UAV can use the links of several partitions concurrently, which leads to great improvement in network throughput. If MIMO capacities (number of the available concurrent transmissions) differs from the number of the partitions (n), the difference can be handled by the partition scheduling.
- **Link space improvement.** If UAVs are flying in obstacle-rich environments such as an urban canyon, the relative distance between the UAVs does not directly refer to the link cost. Also, with the link quality improvement strategies such as packet recovery mechanism [25], a simple, distance-based space partition method cannot result in optimal network topology. To embrace these cases, we can design the *space* considering more than the positions, which can result in a more than 3-dimensional space. By formatting the partition vectors for the augmented space, our topology control layer can operate as Algorithm 1.
- n **granularity.** In Section 3.2, we listed some of the n values, based on the regular polyhedrons derived before. If we can design more S with variable n values following the basic rules in Section 3.2, there are more possibilities to control n in detail, which is advantageous according to Section 4.4.
- **Empirical evaluation.** In Section 4, we measured the performance of the network topologies with our own simulator, which is specialized to measure the property of the topology with power consumption. In this paper, we intended to focus on the clarified improvements in terms of the network topology, so we developed a simulator letting the numerical results be directly derived from the resulting network topology. By inserting our intermediate layer in practical network designs, we can evaluate our proposed scheme into a variety of network environments, by actually implementing devices or the public domain network simulator such as Network Simulator 3 (ns-3).
- **Applications.** As our topology control scheme acts as an intermediate layer between the data link layer and the network layer, it can be broadly used in a wireless network domain. In the case of the UAV network, our system can contribute to performance improvements in multi-UAV surveillance [3], where extending the mission time and guaranteeing the quality of the video transmission service are essential. In addition, our topology control can be deployed in the Wireless Sensor Network (WSN) scenario since the WSN nodes commonly have intensive power

constraints. Due to the compatibility of the system, our proposed scheme can be the breakthrough to solve energy-efficiency and performance-degradation problems in wireless network domains.

6. Conclusions

This study proposed a topology control method for energy-efficient UAV networks. Our system acts as an intermediate layer between the network and data link layer, so it is tolerant of any other network environments. By numerical evaluations, we showed that our space partition method highly reduces the power consumption of the end-to-end connections, while maintaining the node degree and the hop count in a proper amount. We hope that our approach based on the space partition inspires UAV network research domains.

Author Contributions: Conceptualization, H.K.; methodology, S.P., H.T.K., and H.K.; software, S.P., H.T.K., and H.K.; validation, S.P., H.T.K., and H.K.; formal analysis, S.P., H.T.K., and H.K.; investigation, S.P., H.T.K., and H.K; resources, S.P. and H.T.K.; data curation, S.P. and H.T.K.; writing–draft preparation, S.P. and H.T.K.; writing–review and editing, S.P., H.T.K., and H.K.; visualization, S.P. and H.T.K; supervision, H.K.; project administration, H.K.; funding acquisition, H.K.

Funding: The authors gratefully acknowledge the support from Nano UAV Intelligence Systems Research Laboratory at Kwangwoon University, originally funded by Defense Acquisition Program Administration (DAPA) and Agency for Defense Development (ADD).

Conflicts of Interest: The authors declare no conflict of interest. The founding sponsors had no role in the design of the study; in the collection, analyses, or interpretation of data; in the writing of the manuscript, or in the decision to publish the result.

Abbreviations

The following abbreviations are used in this manuscript:

UAV	Unmanned Aerial Vehicle
FC	Fully Connected
SMST	Simple Minimum Spanning Tree
SI	Swarm Intelligence

References

1. Hambling, D. *Swarm Troopers: How Small Drones Will Conquer the World*; Archangel Ink: Venice, FL, USA, 2015.
2. Chung, A.Y.; Jung, J.; Kim, K.; Lee, H.K.; Lee, J.; Lee, S.K.; Yoo, S.; Kim, H. Poster: Swarming drones can connect you to the network. In Proceedings of the 13th Annual International Conference on Mobile Systems, Applications, and Services, MobiSys 2015, Florence, Italy, 19–22 May 2015; Association for Computing Machinery, Inc.: New York, NY, USA, 2015; p. 477.
3. Jung, J.; Yoo, S.; La, W.; Lee, D.; Bae, M.; Kim, H. Avss: Airborne video surveillance system. *Sensors* **2018**, *18*, 1939. [CrossRef] [PubMed]
4. Gupta, L.; Jain, R.; Vaszkun, G. Survey of important issues in UAV communication networks. *IEEE Commun. Surv. Tutor.* **2015**, *18*, 1123–1152. [CrossRef]
5. Bekmezci, I.; Sahingoz, O.K.; Temel, Ş. Flying ad-hoc networks (FANETs): A survey. *Ad Hoc Netw.* **2013**, *11*, 1254–1270. [CrossRef]
6. Li, X.; Guo, D.; Yin, H.; Wei, G. Drone-assisted public safety wireless broadband network. In Proceedings of the 2015 IEEE Wireless Communications and Networking Conference Workshops (WCNCW), New Orleans, LA, USA, 9–12 March 2015; IEEE: Piscataway, NJ, USA, 2015; pp. 323–328.
7. Kagawa, T.; Ono, F.; Shan, L.; Takizawa, K.; Miura, R.; Li, H.B.; Kojima, F.; Kato, S. A study on latency-guaranteed multi-hop wireless communication system for control of robots and drones. In Proceedings of the 2017 20th International Symposium on Wireless Personal Multimedia Communications (WPMC); IEEE: Piscataway, NJ, USA, 2017; pp. 417–421.

8. Katila, C.J.; Okolo, B.; Buratti, C.; Verdone, R.; Caire, G. UAV-to-Ground Multi-Hop Communication Using Backpressure and FlashLinQ-Based Algorithms. In Proceedings of the 2018 IEEE 29th Annual International Symposium on Personal, Indoor and Mobile Radio Communications (PIMRC), Bologna, Italy, 9–12 September 2018; IEEE: Piscataway, NJ, USA, 2018; pp. 1179–1184.
9. Ebert, J.P.; Wolisz, A. Combined tuning of RF power and medium access control for WLANs. *Mob. Netw. Appl.* **2001**, *6*, 417–426. [CrossRef]
10. Ebert, J.P.; Stremmel, B.; Wiederhold, E.; Wolisz, A. An energy-efficient power control approach for WLANs. *J. Commun. Netw.* **2000**, *2*, 197–206. [CrossRef]
11. Iwai, K.; Ohnuma, T.; Shigeno, H.; Tanaka, Y. Improving of Fairness by Dynamic Sensitivity Control and Transmission Power Control with Access Point Cooperation in Dense WLAN. In Proceedings of the 2019 16th IEEE Annual Consumer Communications & Networking Conference (CCNC), Vegas, NV, USA, 11–14 January 2019; IEEE: Piscataway, NJ, USA, 2019; pp. 1–4.
12. Lee, W.; Lee, J.; Lee, J.; Kim, K.; Yoo, S.; Park, S.; Kim, H. Ground Control System Based Routing for Reliable and Efficient Multi-Drone Control System. *Appl. Sci.* **2018**, *8*, 2027. [CrossRef]
13. ur Rahman, S.; Kim, G.H.; Cho, Y.Z.; Khan, A. Positioning of UAVs for throughput maximization in software-defined disaster area UAV communication networks. *J. Commun. Netw.* **2018**, *20*, 452–463. [CrossRef]
14. Wang, X.; Yadav, V.; Balakrishnan, S. Cooperative UAV formation flying with obstacle/collision avoidance. *IEEE Trans. Control. Syst. Technol.* **2007**, *15*, 672–679. [CrossRef]
15. Park, S.; Kim, K.; Kim, H.; Kim, H. Formation control algorithm of multi-UAV-based network infrastructure. *Appl. Sci.* **2018**, *8*, 1740. [CrossRef]
16. Sabino, S.; Horta, N.; Grilo, A. Centralized unmanned aerial vehicle mesh network placement scheme: A multi-objective evolutionary algorithm approach. *Sensors* **2018**, *18*, 4387. [CrossRef] [PubMed]
17. Zeng, Y.; Zhang, R.; Lim, T.J. Wireless communications with unmanned aerial vehicles: Opportunities and challenges. *IEEE Commun. Mag.* **2016**, *54*, 36–42. [CrossRef]
18. Monks, J.P.; Bharghavan, V.; Hwu, W.M. A power controlled multiple access protocol for wireless packet networks. In Proceedings of the IEEE INFOCOM 2001, Conference on Computer Communications, Twentieth Annual Joint Conference of the IEEE Computer and Communications Society (Cat. No. 01CH37213), Anchorage, AK, USA, 22–26 April 2001; IEEE: Piscataway, NJ, USA, 2001; Volume 1, pp. 219–228.
19. Kim, J.; Jeong, J. Range-adaptive wireless power transfer using multiloop and tunable matching techniques. *IEEE Trans. Ind. Electron.* **2015**, *62*, 6233–6241. [CrossRef]
20. Zhang, W.; Wang, J.; Han, G.; Zhang, X.; Feng, Y. A cluster sleep-wake scheduling algorithm based on 3D topology control in underwater sensor networks. *Sensors* **2019**, *19*, 156. [CrossRef] [PubMed]
21. Kim, J.; Kwon, Y. 3-dimensional topology control for wireless sensor networks in presence of interference. In Proceedings of the 2010 Digest of Technical Papers International Conference on Consumer Electronics (ICCE), Las Vegas, NV, USA, 9–13 January 2010; IEEE: Piscataway, NJ, USA, 2010; pp. 471–472.
22. Rauch, H.E.; Winarske, T. Neural networks for routing communication traffic. *IEEE Control. Syst. Mag.* **1988**, *8*, 26–31. [CrossRef]
23. Broumi, S.; Bakal, A.; Talea, M.; Smarandache, F.; Vladareanu, L. Applying Dijkstra algorithm for solving neutrosophic shortest path problem. In Proceedings of the 2016 International Conference on Advanced Mechatronic Systems (ICAMechS), Melbourne, Australia, 30 November–3 December 2016; IEEE: Piscataway, NJ, USA, 2016; pp. 412–416.
24. Kumaran, S.; Sankaranarayanan, V. Congestion Free Routing in ad-hoc Networks. *J. Comput. Sci.* **2012**, *8*, 971–977.
25. Samaraweera, N.K.; Fairhurst, G. Reinforcement of TCP error recovery for wireless communication. *ACM SIGCOMM Comput. Commun. Rev.* **1998**, *28*, 30–38. [CrossRef]

MDPI
St. Alban-Anlage 66
4052 Basel
Switzerland
Tel. +41 61 683 77 34
Fax +41 61 302 89 18
www.mdpi.com

Energies Editorial Office
E-mail: energies@mdpi.com
www.mdpi.com/journal/energies

www.ingramcontent.com/pod-product-compliance
Lightning Source LLC
Chambersburg PA
CBHW062104220926
51850CB00010B/2882
* 9 7 8 3 0 3 9 3 6 1 4 8 9 *